All the Best
Love MacLean/05

536/1000

STANDING AGAINST FIRE

Lieutenant Colonel (Retired)
LORNE MacLEAN, OMM, CD

Published by

GENERAL STORE PUBLISHING HOUSE

499 O'Brien Rd., Box 415, Renfrew, Ontario, Canada K7V 4A6
Telephone (613) 432-7697 or 1-800-465-6072
www.gsph.com

ISBN 1-897113-28-5

Printing by Custom Printers of Renfrew Ltd.

Printed and bound in Canada

Cover design, formatting and printing by Custom Printers of Renfrew Ltd.

No part of this book may be reproduced, stored in a retrieval system or transmitted in any form or by any means, without the prior written permission of the publisher or, in case of photocopying or other reprographic copying, a licence from Access Copyright (Canadian Copyright Licensing Agency), 1 Yonge Street, Suite 1900, Toronto, Ontario, M5E 1E5.

Library and Archives Canada Cataloguing in Publication Data

MacLean, Lieutenant Colonel (Retired) Lorne, OMM, CD, 1951-1987
    Standing against fire : a history of the Fire Service of Canada's military forces and Department of National Defence / Lorne MacLean.

Includes bibliographical references and index.

ISBN 1-897113-28-5

    1. Canada. Canadian Armed Forces. Canadian Forces Fire Service--History. 2. Canada--Armed Forces--Firemen--History. 3. Fire extinction--Canada--History. I. Title.

UC425.M33 2005   355.3'4   C2005-903978-7

A History of the Fire Service of Canada's Military Forces

and

Department of National Defence

By

Lieutenant Colonel (Retired) Lorne MacLean, OMM, CD

Standing Against Fire

Saint Florian
Patron Saint of Firefighters

A History of
The Fire Service of Canada's Military Forces and Department of National Defence

Saint Florian
Patron Saint of Firefighters

Florian was born in Austria about A.D. 250. As a member of the Roman army, he held the rank of field commander in the province of Bavaria in A.D. 304. The Roman Emperor ordered the extermination of Christians, but since Florian was a Christian, he refused to carry out these orders. After hearing this, Aquilinus, who had brought the Emperor's orders, ordered that Florian should be put to death by fire.

Florian told the soldiery that he would ascend the flames on a flight of stairs to heaven. Aquilinus, afraid of Florian's magic, ordered him to be dumped into the Enns River. It was said that as he was thrown from the bridge with a stone tied around his neck, Florian cried, "Lord Jesus, receive my soul," and that an eagle hovered over the spot, casting a shadow in the form of a cross until a Christian lady named Valerie recovered the body and transported it to Linz, where eventually the Austrian monastery of Saint Florian was founded, and where his grave has been the scene of many miracles.

Saint Florian is honoured on the fourth of May and is held as patron of the people of Poland, Upper Austria and the town of Linz. As well, Florian is said to have once stopped an entire town from burning by throwing a single bucket of water onto the fire and, consequently, was adopted as the patron saint of firefighters.

The Maltese Cross—A Symbol of Protection and Honour

Long recognized as the universal symbol of the fire service, the origin of the Maltese Cross is shrouded in antiquity.

During the time of the Crusades, a band of the Knights of Saint John fought the Saracens for possession of the Holy Land, and while doing so they encountered the Saracen's new weapon—FIRE. As the Crusaders advanced on the walls of an enemy fortification, they were struck by glass bombs containing naphtha. When they became saturated with the highly flammable liquid, the Saracens hurled a flaming torch into their midst. Many knights were burnt alive or sustained savage burns to their bodies. Others risked their lives to save their comrades from a painful death. The efforts of these brave men was recognized by their fellow Crusaders who awarded each of these selfless heroes a badge of honour, a cross, similar to the one worn by today's firefighters.

As the Knights of Saint John were from Malta, a small island located in the Mediterranean, the cross that they so proudly wore became known as the Maltese Cross. From that time, the Maltese Cross has been known as a symbol of protection. All who saw it were aware that its bearer was willing to lay down his life in order to protect his fellow man. This symbolism is the reason for its adoption by the fire service and is as representative today as it was in the Middle Ages. The Maltese Cross is the firefighters' badge of honour and signifies that they work with courage to safeguard those whom they have sworn to protect.

Table of Contents

Acknowledgements xxii

Introduction xxv

Chapter I—Organizational Evolution 1
Beginnings of a Fire Service 1
    Overview 1
Years Between World Wars I and II 2
    Benign Decline 2
    The Political Outlook 3
    The Heightening Threat of War 3
Commonwealth Flying Training Plan 3
    Overview 3
    The Need for Fire Service Support 4
Evolving—The Early Years 4
    Operations Principles 4
    Apparatus 5
    The Principal Problems 5
    Advances in Materials and Equipment 5
War Services Fire Protections Committee 5
    Formation and Span of Authority 5
    Evolution into Other Similar Organizations 5
The Corps of Canadian Firefighters 6
    Formation and Administration 6
    Scope of Operations 6
    Achievements and Recognition 6
    In Remembrance 7
    Belated Action—Canadian Government 9
    Interment 9
World War II 10
    Shortage of Manpower 10
    Unprecedented Expansion of the Military 10
    The Disturbing Fire Threat 10
    The Need for a Firefighter Trade 11

British Commonwealth Air Training Plan 11
    Overview 11
    Developing Fire Service Support 12
Joint Services Fire Committee 13
    Structure 13
    Role and Philosophy 14
    Period of Office 14
Integration—Unification—Integration 14
    Overview 14
    Headquarters Formations 14
    The Regional Concept 15
    Fire Protection Philosophy 15
    Equipment Inventories 16
    Facts and Emotions 16
    Firefighter Turmoil 17
    Collateral Damage 17
    Titles and Insignia of Ranks 18
    Bonding 18
    Integration and Unification in Review 18
Staffing Crisis 20
    How it Developed 20
    The Effects 20

Chapter II—Navy Fire Service 27
World War I 27
    Firefighting Force 27
    Protection by Municipalities 27
Between World Wars I and II 28
    Limited Resources 28
World War II 28
    Organizing and Evolving 28
    Laying the Foundation 29
    Firefighter Recruiting 29
    Conditions of Employment 30
    Firefighter Requirements Reassessed 30
    Expansion and Training 30

The Esquimalt Fire School	31
Navy Fire Stations	31
Waterborne Units	32
Aggressive Shopping	32
Harbour Traffic and the Japanese Threat	32

Post World War II ... 32
    Restructuring ... 32
    Changes to the Waterborne Component ... 34
    Bill Fitzpatrick Memorial ... 35

Chapter III—Army Fire Service ... 37
    The Early Times ... 37
World War I ... 37
    Zealous Enforcement ... 37
    Firefighting ... 38
Between World Wars I and II ... 38
    Budgetary Restrictions Dominate ... 38
    Firefighting Resources ... 38
World War II ... 39
    Fire Protection—Starting from Scratch ... 39
    Initial Headquarters Organization ... 39
    Developing a Firefighting Force ... 39
    Manpower Establishments ... 39
    Reorganization at Army Headquarters ... 40
    Unique Characteristics ... 40
    The Fire Marshal's Office ... 40
The North Atlantic Treaty Organization ... 41
    Addressing the Commitment ... 41
    Army NATO Units ... 41
    A Novel Approach Fort Prince of Wales ... 42
    The Fire Service in West Germany ... 42
    Apparatus ... 42
    Mutual Aid ... 43
    Murphy's Law ... 43
    Murphy's Law—Case One ... 43
    Murphy's Law—Case Two ... 44
    The Real Picture ... 44

Chapter IV—Air Force Fire Service 45
World War I 45
    The Beginnings of a Fire Service 45
Between World Wars I and II 46
    Carrying On 46
    Dramatic Development—A Separate Entity 46
    Fire Protection—The Early Years 46
    Appointing a Fire Chief—A Unique Practice 46
    Large Fires—Little Impact 47
    Fundamental Changes 47
World War II 48
    Setting Up a Fire Service Organization 48
    The Japanese Threat 48
    Organizational Change 49
    A Sidebar of Interest 49
Post World War II 49
    Exodus of Personnel 49
    The Struggle for Leadership 50
    At the Fire Departments 50
The Cold War 51
    International Tension 51
    Materiel 51
    Leadership Foundation 51
    Bilateral and Multilateral Agreements 52
NATO 52
    Composition of the Air Force in Europe 52
    Profile of the Fire Service 53
    Equipment 53
    Unique Operations Demands 53
    New Firefighting Technique 54
NORAD 54
    Purpose and General Structure 54
    USAF Strategic Air Command Bases 55
    USAF Strategic Air Command Bases—Staffing 55
    Strategic Air Command Bases—Apparatus 56
    USAF Strategic Air Command Bases—Termination of Agreement 56

    Nuclear Munitions 56
        Why and How 56
        Nuclear-Capable Sites 57
        Physical Security 58
        Operations—Restructuring to Meet the Need 58
        New Responsibilities—New Language 58
        Firefighting and Nuclear-Hazard Considerations 59
        Standard of Performance 59
        Far-Reaching Effects on the Fire Service 60
        Training to Meet the Demands 60
        The Nuclear-Training Dividend 61
Miscellaneous Commitments 61
        Aircraft Arrestor Engines 61
        UN/NATO/NORAD Missions 62
Summary 62
Escalation—Steady State—Decline 62
Tales for the Telling 64
        Overview 64
        Embarrassment Personified 64
        Over the Hill 64
        Grostenquin in the Raw 65
        Marvellous Marville 67
        Where the Dust Settled in Metz 69

Chapter V—Personnel 71
        Overview 71
        Conditions of Employment 72
        Advancing with the Times 72
        Changes—Fire Department Rank Structure 72
Physical Fitness 72
        Background 72
        First FFSPFP 72
        Training Practices 73
        Evaluation Standard 73
        Administrative Hurdles 74
        National Policy 74

Complaints—First FFSPFP	74
Complaints Answered	75
Forces-Wide Physical Fitness Program	75
CSTF	76
Research—Minimum Physical Fitness Standards for the Forces	76
Looking Back and Ahead	77
New Research— Firefighter Physical Fitness Program	77
Second FFSPFP	77
Implementation	79
Complaints—Second Firefighter Specific Physical Fitness Program	79
Maintaining a Fire Service Officer Corps	**79**
Commissioning from the Ranks Plan	79
Special Commissioning Plan	80
Special Ranks Commissioning Plan	80
University Training Plan—Non-Commissioned Members	80
Service in the Broader Sense	**80**
Introduction	80
The Council of Canadian Fire Marshals and Fire Commissioners	81
Executive Positions	81
The Canadian Association of Fire Chiefs—Mission	81
The Canadian Association of Fire Chiefs—Profile	81
Involvement—Fire Service Members	82
The Office of the President	82
Fire Prevention Canada	82
Other National and International Organizations	82
Sidebar—The Fire Service Exemplary Service Medal	83
Honours and Awards	**84**
Introduction	84
Air Command Commander's Commendation	84
British Empire Medal	85
Certificate of Meritorious Conduct	88
Chief of Defence Staff Commendation	88
Fighter Group Commander's Merit Award	88
George Medal	88
Medal of Bravery	89
Order of Military Merit	90
Queen's Commendation for Brave Conduct	90

xiii

Fatalities—Line of Duty ... 91
    Introduction ... 91
    The Honour Roll ... 91
Burnishing the Ties that Bind ... 91
    Overview ... 91
    Box 1701 Club ... 91
    Curling Bonspiel ... 92
    Golf Tournament ... 94
    Western Canada Reunions ... 94
    Eastern Canada Reunions ... 94
    The Round-Up ... 95
    The Fire Service On Line ... 95

Chapter VI—Professional Development ... 97
    The First Schools of Firefighting ... 97
    Critical Shortage of Apparatus ... 97
    Employment of Course Graduates ... 98
    Addressing an Urgent Need for Firefighters ... 98
    The Various Locations of the Fire Schools ... 99
    Establishing and Stabilizing at Borden ... 99
    Change and Stability ... 99
    Updating and Broadening the Professional Development Program ... 99
    Tragedy Strikes ... 99
    Training Evolves and Continues ... 99
    Students to be Demonstrably Physically Fit ... 100
    The Transition into an Academy ... 101
    Getting the Academy Up and Running ... 102
    Background—Academy Infrastructure ... 103
    Developing New Facilities ... 103
    A Catalogue of Facilities ... 104
    Keeping in Step with the Times ... 105
    Familiarization Training—New Military Engineer Officers ... 105
    International Recognition—The CFFA Teaching Institution ... 105
    Certification of Individual Qualifications ... 106
    Training Foreign Students ... 106
    CFFA Marketing ... 106

    The Academy in Profile 106
    The Apprentice Course 107
    On-Job Training 107
    The Journeyman Course 107
    The Supervisor/Commander Course 107
    The Manager/Commander Course 107
    The Fire Investigation Course 108
    The Fire-Prevention and Life-Safety Course 108
    The Hazardous Materials Technician Course 108
    The Hazardous Materials Manager / Incident Commander Course 108
    Professional Development of Commissioned Fire Officers, 1952–69 108
    Professional Development of Commissioned Fire Officers, 1970 Onwards 108
    Professional Development of Commissioned Fire Officers, Fire Service Specific ... 109
    Summary, Professional Development of Commissioned Fire Officers 109
    Recertification Training 109
    Distance Learning 109
    A Film and a Story 110
    Events of Special Interest 112
    Summary 112
Picture Gallery—VI 113

Chapter VII—Fire Prevention 121
General 121
    Philosophy and Structure 121
    Raising the Quality Standard 122
The Three "Es" of Fire Prevention 122
    Overview 122
Fire Loss-Limiting Engineering 123
    The Beginnings 123
    Evaluating Construction Projects—A Structured Method 124
    The Origin of Sprinkler Systems 124
    Automatic Sprinkler Systems 124
    Foam-Deluge Systems in Aircraft Hangars 124
    Carbon Dioxide Fixed-Pipe Systems 125
    Halogenated Fire-Extinguishing Agents 125
    Canada's Military Embraces Halon 126

Demands for Halon Fire-Suppression Systems	126
Halon—Waxing and Waning	127
Halon—Putting the Genie Back in the Bottle	127
Halon—A Summary	128
Fire Safety Education	129
Evidence of the Need	129
Program Details	129
Enforcement	130
Observing and Acting	130
Fire Investigation	130

Chapter VIII—Materiel	131
Personal Protection	131
Turn-Out Gear	131
Smoke-Eaters and Breathing Apparatus	132
The Chemox Self-Contained Breathing Apparatus	133
Compressed Air Debuts	133
The Nomex Work Dress	134
Breathable Air—Filling Station	134
Air Carts	134
Portable Air Supply System	135
Rescue and Extraction Equipment	135
Overview	135
Air-Bag Lifting System	135
Air Chisel	135
Hurst Jaws of Life	136
Power Hawk P-16 Rescue System	136
Fire Suppression	136
Asbestos Fire Blankets	136
Water Fog	136
Firefighting Foams	137
General Characteristics of Foam	137
Establishing and Maintaining a Foam Blanket	138
Protein Base Foam Concentrate	138
Dry Chemical	138
AFFF	139

MXF	139
Thermal-Imaging Cameras	140
Summary	140
Picture Gallery VIII	**141**
Introduction	141

Chapter IX—Apparatus ... 149
Background ... 149
- Overview ... 149
- Romans and Greeks Lay the Groundwork ... 149
- Following the Footsteps of the Romans ... 150

Evolution in Modern Times ... 150
- The Immense Impact of World War II ... 150
- Early Development—Aircraft Rescue and Firefighting Vehicles ... 150
- Experimenting with Dry Chemical ... 151
- The Impetus for Change ... 151
- A Serious Shortcoming Finally Addressed ... 152
- Introduction of DC ... 152
- The First DC Crash Truck ... 152

International Standards ... 153
- ARFF—Calculating Requirements ... 153
- Calculating Response Times ... 153
- Satisfying Vehicular Requirements in Principle ... 153
- Minimum Number of ARFF Vehicles ... 154

The Roster ... 154
- Introduction ... 154
- 1920 The First Canadian Fire Truck ... 154
- The Nashwack Fire Boat ... 154
- 1938 Ford Crash Tender ... 154
- 1939 GMC/American LaFrance Pumper ... 154
- 1939 International Pumper ... 154
- 1941 International Bickle-Seagrave Pumper ... 155
- 1941 Code 30 (Redesignated G10) Pumper ... 155
- 1941 Code 33 (Redesignated G15) Crash Tender ... 155
- 1942 American-LaFrance Pumper ... 156
- 1944 American-LaFrance Aerial ... 156

1945 Gifts from the United States	156
1948 G17 DC Crash Vehicle	156
1951 Chevrolet Range Truck	156
1951 G18 DC Crash Vehicle	156
1951 G9 Pumper	156
1952 Bickle-Seagrave Aerial	157
1952 Ford Crash Tender	157
1952 Pumper	157
1952 G11 Triple Combination Pumper	157
1952 G13 LRV	157
1953 GMC-Bickle-Seagrave Pumper	158
1953 G21 Thornycroft MFV	158
1953 A21 FWD MFV	159
1954 G23 MFV	159
1954 GMC-Bickle-Seagrave Pumper	160
1954 FWD Pumper	160
1954 Bickle-Seagrave Pumper	160
1954 Walter-Bickle-Seagrave MFV	160
1955 International-Thibault Pumper	160
1956 International Harvester Crash Truck	160
1957 G19 Alvis MFV	160
1957 G8 Pumper	161
1958 The High-Pressure Pump	162
1958 Ford Range Truck	162
1958 E62 Street Flusher	162
1960 Fargo Aerial	163
1960 Tracked DC Crash Vehicle	163
1962 Ford Range Truck	163
1963 MPV Pumper	163
1964 Thibault Pumper	163
1964 G19 Sicard MFV	163
1965 Ford, King-Seagrave Pumper	164
1965 Walter, King-Seagrave MFV	164
1966 Thibault Pumper	164
1969 Heliport ARFF Vehicle	164
1970 UNIMOG Quick Response Vehicle	164

1972 International-Pierreville Pumper 164
1972 Oshkosh M-1000 MFV 164
1972 Oshkosh M-100 Sidebar 165
1972 Oshkosh T-1000 Major Foam Vehicle 165
1974 MAN Gebruder Bachert 165
1974 Ford Saskatoon Barton-American Pumper 165
1975 International King-Seagrave Pumper 166
1975 Ford Aerial 166
1976 FLEXTRAC DC Crash Vehicle 166
1997 Kenworth Pierre Thibaullt Aerial 166
1978 Scot, Pierre Thibault Aerial 166
1979 IHC Pierre Thibault Pumper 166
1979 Oshkosh M-1000 MFV 166
1981 International Pierreville Aerial 166
Increased Demands—New Solutions 166
1981 RIV 166
1981 CDN Research MFV 167
1982 Ford King-Seagrave Pumper 167
1982 MPV Range Truck 167
1982 Walter 4500 MFV 167
1982 Universal Go-Track Crash Vehicle 168
1984 Western Star Range Truck 168
1984 Ford Thibault Pumper 168
1985 Walter 4500 MFV 168
1987 E-ONE TITAN MFV 168
1988 Ford Superior Pumper 168
1988 Unimog Range Truck 168
1989 Timberjack Range Truck 169
1989 Waltek Universal Go-Track Crash Truck 169
1993 Spartan-Thibault Pumper 169
1993 Oshkosh 6000 MFV 169
1995 Navistar-International Range Truck 169
1995 Oshkosh T-1000 MFV 169
1995 Emergency One P-150 170
1995 Tibotrac Water Tower/Pumper 170
1995 Oshkosh-T-1000C MFV 170

    1996 Ford Rescue Van 170
    1997 Ford Rescue/Hazmat Van 170
    1997 International Navistar/Eastway-Paystar 5000 Water Tanker 170
    1997 Ford F150 Davtair DA-1000 170
    2000 E-ONE Cyclone Platform 171
    2000 E-ONE Aerial 171
    2000 Walter P4-6000 MFV 171
    2001 E-ONE Typhoon Pumper 171
    2002 Western Star Range Truck 171
Picture Gallery IX 172

Chapter X—Operations 223
    Introduction 223
    1943—Fire—Shipboard, Halifax Harbour, Nova Scotia 223
    1945—Fire—Bedford Magazine, Bedford, Nova Scotia 223
    1955—Crash—York Freighter, Edmonton, Alberta 225
    1955—Rescue—Crash, Bristol Freighter, Marville, France 225
    1955—Fire— Aircraft Hangar, Langar, England 226
    1956—Fire—Montmendy, France 226
    1958—Crash—AVRO Arrow, Toronto 228
    1963—Thoughtful Experiment 229
    1966—Crash—CF101 Voodoo, Chatham 230
    1967—Crash—Air Canada DC 8, Ottawa 231
    1967—Fire—Motor Transport Building, Ottawa, ... 232
    1973—Rescue—Cold Lake, Alberta 233
    1974—Fire—Fuel Farm, Courtney, British Columbia 234
    1976—Rescue—McIvor Lake, British Columbia 235
    1977—Crash—Argus, Summerside, Prince Edward Island 236
    1982—Fire—Use of MEF on a Building Fire, Lahr, West Germany 241
    1982—Crash—Hercules, LAPES Incident, Edmonton 242
    1984—Fire—Aircraft Hangar, Baden, West Germany 243
    1985—Crash—Hercules Mid-Air Collision, Edmonton, Alberta 243
    1988—Fire—Barge FOSS 290, Comox, British Columbia 247
    1993—Fire—City of Lahr, West Germany 248
    1997—Fire—Dartmouth, Nova Scotia 249
    1997—Fire—The Gijon Fishing Trawler, Esquimalt, British Columbia 251

Summary 251

Bibliography 253

Appendices 255
    Appendix A—Key Appointments
    Appendix B—Descriptions & Definitions
    Appendix C—Acronyms & Abbreviations
    Appendix D—History Teams
    Appendix E —Radar Stations
    Appendix F—Firefighters—Minimum ARFF Standards
    Appendix G—Calculating Minimum ARFF Requirements
    Appendix H—Operations Assignments

ACKNOWLEDGEMENTS

Research and Development

This project was pursued with sponsorship support from Lieutenant Colonel Marc Desjardins, the CFFM, and in the broader sense, the CFFM's office as a whole. In addition to those mentioned in the introduction, the members of the teams shown in Appendix D deserve special credit. They were named without as much as a by-your-leave and in a reassuring demonstration of cohesiveness and esprit de corps, all, save two of 45 selected, participated with enthusiasm.

Some individuals deserve special recognition. Specifically: Lieutenant Colonel (Retired Military Engineer) Ken Holmes who used his extensive experience in researching archived material and providing guidance for others to work in this field; Chief Warrant Officer Steve Shand and Major (Retired) Don Carmichael, for their tireless efforts; and Chief Warrant Officer (Retired) Joe Walker who unfailingly responded to my endless requests for information, often with the active support of the other members of the www.Firehouse651.com Web-site management team, namely Captain Al Rau, Master Warrant Officer (Retired) Paul Landry and Master Warrant Officer (Retired) Don Teed. Thanks also to Captain Mike Blow who managed the project's meagre financial resources.

Finally, special thanks to Ms. Jane Munro and to Ms. Susan Code McDougall, both of whom used an eagle eye and gentle critiques to nudge me, perhaps with little avail, towards the use of one of Canada's two official languages, rather than one I had invented for my own private use.

Donors

Financial Support
Critically important financial support was provided by:

Corporate
- FTS Fire Training Systems
- George Cowan Enterprises
- Levitt Safety
- MSA
- National Fire Protection Association

Private
- Lieutenant Colonel (Retired) L. MacLean
- Lieutenant Colonel (Retired) H. Singleton

The RCN 1000 Pumper is a history in itself. It entered service at HMCS Halifax Dockyard in 1942, and in 1949, it was moved to the Naval Armament Detachment (NAD) in Dartmouth. On the evening of its arrival, FSO 1 Bernie Levangie was giving NAD firefighters driver training when the vehicle jumped a ditch and hit the foundation of Building W-21, smashing the front bumper, the grill, the top engine hood/nose section, both fenders and the radiator. This accident marked the end of the active career of RCN 1000 in Canada's military.

Shortly thereafter, it was sold to Lawrencetown Fire Department in Annapolis Valley, NS, and brothers Merle and Cecil Slauenwhite, both mechanics, arrived at the Dockyard with a few tools and a radiator from an old farm tractor with the aim of making it sufficiently road-worthy to drive to Lawerencetown. Within one half day they had it rolling. As Pumper No. 5, it served the community for many years.

Following a period of neglect in pumper purgatory, in this case a farmer's field, it was spotted by Deputy Fire Chief Gerald Geddes of the Dartmouth Detachment in September 1990.

Subsequently, it was rescued by Dockyard Fire Department and lovingly nursed back to the pleasing state it (and us) now enjoy.

INTRODUCTION

This book chronicles the history of the Fire Service that has provided protection against the ravages of fire for the personnel and material resources of Canada's military services. The span of history covered begins in the era of the First World War and continues to 2004.

For a number of reasons, the history of the Fire Service is a complex topic. Primary amongst these reasons are the span of time covered; the many and varied peaks and valleys that have been the hallmark of Canada's military; the multifaceted roles of the Fire Service as executed by our firefighters, be they military or paramilitary civilians; and, most important of all, the lack of any deliberately retained historical records.

The role of the Fire Service is to minimize the adverse effects of fires, accidents and aircraft crashes that occur as a result of peacetime activities, or overt or covert actions by an enemy on the capability of the Canadian Armed Forces to conduct military operations. Inherent in this role are multiple spheres of responsibility, in that it requires the Fire Service to be one of the very few fire service organizations that covers all four factors of fire protection, namely: fire loss-limiting engineering, fire-safety education, enforcement of fire-safety regulations and standards, and firefighting and associated rescue operations.

While a fire service within the Canadian military has been a reality since the First World War, documentation describing early operations is sketchy and very difficult to chronicle with the degree of accuracy and detail they rightfully deserve. This is particularly true of its formative years. While much care has been taken to be as historically correct as possible, some inferences had to be drawn and opinions expressed, as a means of capturing the flavour of the time, as is common to any recording of history.

From its humble beginnings, featuring poor equipment, poorly trained personnel and meagre resources, the Fire Service evolved and upgraded itself to the point where, for well in excess of half a century, it has been a truly professional entity. This fact has been vividly and repeatedly demonstrated, as is reflected in this document. For quite some time, individual firefighters have been qualified to internationally recognized standards of proficiency, credit for which is due to the tireless efforts of dedicated firefighters of all ranks who have worked diligently throughout the years to improve the firefighters' professional qualifications and enhance the image of the Fire Service.

The readers who served through the events described in this record may well find cause to challenge some of the details contained herein. To them, we state that some material must be viewed with the realization that there is limited documentation left to describe the circumstances. With the foregoing in mind and in an effort to avoid unnecessary controversy, efforts have been taken to avoid the indiscriminate use of individuals' names, except where they are inextricably entwined with the actual event being described.

The earlier efforts put forth to prevent our history from becoming ever more misty through the passage of time are acknowledged. Amongst these efforts were those of Mr. Gordon Lay regarding the Naval Fire Service, Major Phil Brown's introspective narrative of his experiences in the wartime Royal Canadian Air Force (RCAF) as a pilot and later as a fire-protection officer, and Flight Lieutenant John Cowell's account describing some of the significant milestones relative to the Fire Service. Several years later, in the late 1980s, Captain George Cowan blended the earlier documents and did further research on the project. The ensuing package was taken in hand by Captain Evan Evans who moved it further along in the early 1990s.

Finally, in late 2002, Lieutenant Colonel Marc Desjardins, Canadian Forces Fire Marshal (CFFM), approached me with the suggestion that, if we were to avoid losing our roots, any further delay in compiling our history would be a critical risk, surely an inarguable view. After some discussion it was decided to activate a project, with sponsorship support by the CFFM, with the specific aim of pursuing the task to its completion, using the work that had previously been done as the starting point. As the project got underway, it quickly became evident that there was a great deal of work to be done. Now, some two and a half years later, this impression can be recognized as an irrefutable fact.

Looking back at the profile of Canada's military services over the better part of a century is at times a pleasurable thing to do while at other times quite painful. Essentially, it depends on what time frame one focuses on. Not surprisingly, the highs and lows of the Fire Service have been in lockstep with the fortunes of the military as a whole. Fortunately, one of the more laudable features of the human mind is its propensity to retain pleasant memories with greater clarity than is the case with unpleasantries.

To conjure up memories of what has taken place over more recent years regarding the alarming erosion of Canada's Armed Forces is painful to contemplate. Fortunately, there have been more heady times which, in the main, can be traced to three major events: Canada signing the treaty that created the North Atlantic Treaty Organization (NATO) in 1949, the Korean War in 1950, and the North American Aerospace Defence Agreement (NORAD) between Canada and the United States in 1958. As things turned out, NATO and NORAD were to have far-reaching impacts on the Fire Service, imposing a wide range of commitments in Europe as well as in Canada, some never before experienced.

To execute these new responsibilities at the near error-free standard demanded for effective wartime operations required a great deal of practice. Frequent exercises and evaluations were conducted by

Headquarters Formations. On completion of such an evaluation, if the base/wing/station passed, it was common practice for the unit commander to declare a "stand-down" of at least one day in recognition of the many extra hours, both daytime and nighttime, expended to achieve the demanding level of proficiency that was required. Quite often this would result in a Friday, Saturday, Sunday long weekend. However, firefighters could not take this break to enjoy their well-earned laurels because it would be in violation of the fundamental principle of "Standing Against Fire." For the Fire Service, a period of non-vigilance would be totally incompatible with this principle. It is one of the prime reasons we have chosen this phrase as the title for this book.

Following in the footsteps of many others in the process of obtaining and compiling facts has been a demanding undertaking, but pales in comparison to what was required to interpret the implications of any given piece of information. In doing this, the overall aim was twofold in nature: firstly to ensure our proud history did not disappear into the mists of time; and, secondly, to illuminate the enviable spirit and lore of the Fire Service.

I enjoyed a great deal of assistance in gathering the information contained in this book, for which I am most grateful. This notwithstanding, the responsibility for the interpretation of any fact, material, or situation rests solely with me.

It is my fervent hope that all members will enjoy this record and will find that the dual aim has been achieved.

Lieutenant Colonel (Retired) Lorne MacLean, OMM, CD

CHAPTER I

ORGANIZATIONAL EVOLUTION

The BEGINNINGS OF A FIRE SERVICE

Overview

The beginnings of the Fire Service protecting the resources of Canada's military services are difficult to determine. For the Navy, shore facilities were protected in the same manner as ships. This took the form of an officer-of-the-day (watch) supported by duty-watch personnel. However, because the two main dockyards, one in Esquimalt and the other in Halifax, were basically manned by a civilian work force, the duty-watch personnel came from docked naval ships. Senior management at the dockyard invariably consisted of Royal Canadian Navy (RCN) officers; therefore, the practice of appointing an officer-of-the-day was universally established.[1] For the naval installations at Halifax and Esquimalt, the City of Halifax and the City of Victoria Fire Departments provided additional protection. It was not until 1942 that a land-based Naval Fire Service began and eventually accepted complete responsibility for dockside fire protection.

For the Army, the principles outlined in the British Army's *Fire Manual for Hutment Camps* and the *Army Fire Manual* were followed. A 1916 edition of these manuals placed emphasis on "discipline, order, the proper training of pickets and constant patrol inspections and the complete readiness of equipment."

An event that would have a pivotal effect on the development of a fire service occurred on a wintry day

---

[1] Lieutenant Commander Gordon Lay, handwritten recollections (unpublished, May 1963).

in 1917, when Lieutenant Colonel (later Brigadier General) Cuthbert G. Hoare of the Royal Flying Corps selected Camp Borden, Ontario, as the site for flying operations in Canada.

The first phase of the history of the military firefighter has a direct association with this event. This site was eventually to become the senior station of the RCAF.[2] The camp's isolated location and the need to be self-reliant with regard to fire protection may have been instrumental in stimulating the beginnings of a Canadian military fire service that, initially, comprised two distinct and separate functions: aircraft crash responses and structural firefighting.

Although written records of fire services during the First World War are scarce, documentation confirms the existence of an organized military fire brigade. A photograph (circa 1918) of uniformed military firefighters illustrates personnel on duty at Camp Borden.[3]

Throughout the history of firefighting in all three services, success at fire prevention often seems to have delayed the introduction of new programs to upgrade the capabilities of the Fire Service. Firefighting and fire protection needs were recognized only when a major fire or other disaster occurred.

The YEARS BETWEEN WORLD WARS I AND II

Benign Decline
The period between World War I and World War II passed with little noticeable changes or improvement to the military fire service. This was understandable considering the severe reductions experienced by the armed forces as a whole. Manpower and funds were reduced to the point that the very existence of the military forces seemed threatened. Henri Beland, who as Minister of Health in Prime Minister Mackenzie King's cabinet during the years 1921–25, was charged with the responsibility for reinstating soldiers into civilian life. He was a strong proponent for reducing expenditures for the military. Typical of his approach, he expressed misgivings respecting the commercial and scientific applications of aviation in Canada[4] and voiced his determination not to vote any money for military purposes by declaring, "the war is over and the Government should put a stop to these expenditures."[5] Prime Minister Mackenzie King agreed. "To enter this year," he remarked, "before the work of demobilization is completed, upon an air service for military purposes is the height of absurdity . . . We can well afford to dispense with the military end of it this year, and I think that we shall reduce the vote for air services by $800,000."[6] This proposal carried in the House of Commons by a vote of 46 to 26.[7]

The RCAF,[8] newly formed with an establishment of only 68 officers and 307 men, spread from Dartmouth to Vancouver, and having to define its role as an important military arm in an era of peace, was greatly affected by this decision.

With the exception of new aircraft procurement, new programs were cut in an effort to maintain the viability of the remaining resources. The Army and Navy also

---

2   S.F. Wise, *Canadian Airmen and the First World War* (Toronto: University of Toronto, 1980).
3   *The Montreal Standard*, 26 October 1918.
4   Henri Beland was appointed to the Senate in 1925.
5   James Eayers, *In Defence of Canada* (Toronto: University of Toronto Press, 1964), p. 200.
6   Ibid., p. 200.
7   Ibid., p. 201.
8   Douglas, *The Creation of a National Air Force*, p. 62.

suffered, and they had to stretch their meagre financial allotments. The political will to maintain the armed forces at a sizeable level of manpower did not exist during this period.

The Political Outlook

Politicians of the time had readily rallied around the belief that Canada had taken part in a victorious crusade to win the war to end all wars and was, therefore, justified in settling down to a well-deserved period of peace. "There is no world menace," declared the leader of the Opposition during a 1920 House of Commons debate on military expenditures.[9] It was a cry heard often among politicians of all parties immediately after the First World War.

The Heightening Threat of War

As the intervening years slipped by, a heightened threat of war became evident. An awareness of the potential for losses of important war material through fire prompted a re-examination of the strategic approach to military fire services. With civilian fire authorities forming the vanguard for change, a new sensitivity towards fire protection emerged and an in-depth reappraisal of military fire protection services was justified.

COMMONWEALTH FLYING TRAINING PLAN

Overview

The intention was to train the Commonwealth pilots who were needed for duty in the European theatre. Pilots trained at Borden ultimately fought in an action that was taking place over the World War I battlefields of Europe, Asia, and Africa and above the adjacent waters.

The Royal Naval Air Service (RNAS) and the Royal Flying Corps (RFC) were joint developers of Camp Borden and had a vested interest in protecting the expensive infrastructure. Development of the fire brigade organization at Camp Borden during the war years was no doubt given impetus by the presence of many aircraft and the highly flammable fuel that powered them.

Aircraft hangars were constructed within six weeks of the RFC's Lieutenant Colonel Hoare's initial inspection. Personnel quarters were simultaneously built with sufficient accommodation to permit a training squadron to form and start assembling aircraft.[10] In just four months, the contractors erected 57 buildings, cleared and levelled 850 acres / 344 hectares and sowed the ground with grass seed, built five miles / eight kilometres of asphalt road and laid additional sewage pipes and rail sidings. They installed an electrical system and strung telephone lines to connect the field with Toronto and neighbouring towns. So successful was this method of development that it became the construction standard for later fields. So rapid was the camps progress that Hoare, in a report to the War Office, "observed the work appears to be put through at a speed here which is unknown in England."[11]

By war's end, 1918, Camp Borden was recognized by aviation experts as one of the most up-to-date aviation training facilities in the world. This was quite a flattering endorsement considering the speed with which the installation was constructed.[12]

Materials used in building the aircraft hangars and repair buildings were chiefly wood and other combustible material. As the British were more familiar

---

9 James Eayers, *In Defence of Canada*, p. 214.
10 S. F. Wise, *Canadian Airmen and the First World War* (University of Toronto, 1980).
11 Development of the Royal Flying Corps in Canada (memorandum, App.A,), Air 1/721/48/4.
12 Madeleine Stace, *History of the Construction Engineering Branch* (Ottawa: Department of National Defence, 1963), p. 19.

with non-combustible European construction methods, concern over the density and combustibility of the material prompted the RFC to establish an on-site structural fire brigade. These fears were not without substance as, of the original 17 hangars, several were later destroyed by fire. Nevertheless, to the builders' credit, most of these temporary wartime structures were still providing useful service some 75 years later. The Canadian Forces Fire Academy (CFFA) occupied two of the structures until it moved into new accommodations in 1986. Ironically, the specification spelled out by Lieutenant Colonel Hoare was that the buildings at Borden were to be of semi-permanent nature, and he arranged for the cheapest form of construction possible compatible with strength.[13]

The Need for Fire Service Support
Aircraft accident statistics sent by RFC/RAF Canada headquarters to the War Office for the period April 1917 through May 1918 indicate that at Camp Borden there were 49 aircraft accidents and 30 fatalities during that one-year period. The figures clearly indicate the need for aircraft rescue and firefighting services, although records of structural fire occurrences were not located.

The emergence of an organized military fire brigade at Camp Borden was a noteworthy event because it helped set the stage for a fire service that would become an integral part of future military organizations.

EVOLVING—THE EARLY YEARS

Operations Principles
Operationally, there were local unit variations with regard to firefighting response procedures, depending on the experience of the individual in charge of the department. Training of non-professional fire crews was carried out on site or by instruction from local fire brigades. This instruction, no doubt, reflected the particular prejudices of the local trainers with regard to firefighting methods or techniques. However, the roots of the Fire Service would have been vested in the senior firefighters of that time, as the science of fighting fires was not well developed in those days, although the effects of fire were well known. The application of water in copious quantities was the basic principal applied to fighting fires regardless of size. General Duty (GD) personnel were able to quickly learn the rudiments of firefighting, and they became a pool of readily available manpower. The philosophy was to contain fire spread with little appreciation or even the perceived need to take the Fire Service beyond the established level.

Units that operated during these years would have, in all likelihood, carried out their day-to-day operations along comparable lines, depending on their equipment and the local know-how available. Firefighting equipment inventories would have been varied, in some cases having only hose reel carts as their primary response apparatus. Until 1939, only three stations had their own motorized fire vehicles. The remainder "made do."

The using of hose reel carts as potential support for firefighting operations involving mobile fire apparatus remained until the late 1970s.

Early firefighting operations placed considerable reliance on the local municipal fire departments. Slowing the progress of the fire and protecting the immediate exposures became the main objective of the station's fire response. Firefighting vehicles were operated by drivers from the Transportation Section, which continued until the beginning of World War II. The rest of the response crew was drawn from the duty fire pickets. The instruction given was rudimentary,

---

[13] S. F. Wise, *Canadian Airmen and the First World War* (University of Toronto, 1980).

with little resemblance to the thorough training that would be demanded by mid-century and beyond.

Apparatus

Vehicles of 1914–18 vintage and their firefighting packages were primitive by today's standards. Nevertheless, they were capable of pumping a respectable 325–350 G/min / 1475–1590 L/min, using a rotary gear-type pump. These pumps operated on a positive displacement principle that, although effective, occasionally developed serious problems when discharge orifices became obstructed during pumping operations. While there were centrifugal pumps installed in stationary pump houses, these were too large to mount on mobile equipment.[14] Centrifugal pumps did not become the virtually exclusive choice for fire apparatus until the early 1950s.

The Principal Problems

The principal problems facing a fire brigade response crew dispatched to fight a large fire in 1918 arose from marginal or unreliable water supplies. In addition, they had to deal with limited pump capacity and hose lines that were inclined to burst at the most inappropriate time. These problems and a myriad of other equipment difficulties made the chances of successfully extinguishing even a modest fire a formidable task. Eventually, hose standards improved and testing procedures were incorporated into on-site training exercises. Techniques of combatting and controlling fires were largely confined to putting the wet stuff on the red stuff and liberal servings of the time-tested "surround and drown" tactics.

Advances in Materials and Equipment

Continual advances were made in the development of firefighting equipment and extinguishing agents, but some of the more sophisticated fire-suppression chemicals had yet to make their appearance. Nevertheless, the highly efficient (albeit toxic) carbon tetrachloride (CTC), clorobromomethane (CBM),[15] chemical foam, and soda and acid were available, as was carbon dioxide (CO_2). Use of these agents was confined to hand-portable fire extinguishers, with the exception that there was some 40 G / 180 L chemical foam, wheeled units, but it was still water that provided the mainstay for most firefighting operations.

The WAR SERVICES FIRE PROTECTION COMMITTEE

Formation and Span of Authority

The War Services Fire Protection Committee (WSFPC), although respecting the past recommendations made by various civilian fire prevention authorities under the auspices of the Dominion Fire Commissioner, wished to place their own stamp on future fire inspections of military installations. The WSFPC gradually assumed all authority and responsibility for providing technical advice on matters of fire safety throughout the three services. Consolidation of a group of experts to provide direction and ensure continuity among all three services in matters of fire protection eliminated many problems before they could adversely affect operational effectiveness. Eventually, the committee also assumed fire protection responsibilities for the Department of Munitions and Supply, regarding companies manufacturing critical war material.

Evolution into Other Similar Organizations

The WSFPC served early wartime Canada exceedingly well and remained active until their last meeting on May 6, 1942, when the Joint Fire Marshals Committee (JFMC) replaced them. The formation of the WSFPC had provided a catalyst for the formation of the Joint

---

14   Memoirs of Major Walter Sinclair, The RCAF Fire Service, Major W. Sinclair (unpublished handwritten recollections, January 1990).

15   Forerunner and close relative of the halogenated extinguishing agents.

Services Fire Committee (JSFC), a body that served the Canadian Forces Fire Service until integration in 1966, followed by unification in 1968.

The CORPS OF CANADIAN FIREFIGHTERS

Formation and Administration

Following a visit to the United Kingdom (UK) in 1941 by Prime Minster Mackenzie King, Canada agreed to provide a corps of Canadian firefighters to serve in the UK. Four hundred Canadian firefighters volunteered their services to help the people of the UK respond to the menace of Nazi air attacks. These paramilitary civilian firefighters who went overseas had a close affinity with the armed forces. The strongest personal thread of the relationship is that of the commanding officer of the corps; G.E. Huff, MM, had been a flight lieutenant fire prevention officer in the RCAF, at No. 2 Training Command. After several appeals he had obtained his release from the RCAF in order to take command of the corps.[16]

For the purposes of administration, this corps was placed under the direction of the Minister of National War Services. Recruiting began in March 1941, and within a few months, had achieved a total strength of 422 members drawn from 107 municipalities representing all Canadian provinces. The corps membership included: 12 from Alberta; 37 from British Columbia; 29 from Manitoba; 4 from New Brunswick; 6 from Nova Scotia; 285 from Ontario; 3 from Prince Edward Island; 27 from Quebec; and 20 from Saskatchewan. Three used England as their enrolment address. Of these numbers, 406 were sent overseas by December 1941 and five remained at the Corps Headquarters in Ottawa for administration purposes, two of which later left for the UK.

Scope of Operations

Upon arrival in the United Kingdom, the volunteers were given a four-week course in firefighting, rescue work and drill. On completion, they were posted to fire stations located in Southampton, Portsmouth, Plymouth, Bristol and London.

Canadian units attended all fires in localities where they were stationed. Many of these fires were caused by German bomber attacks which added great risk to the responding crews. During bomber raids on London in February 1943, the headquarters unit came into action both in firefighting and rescue operations, and its performance was highly commended. While many members of the corps were injured in the course of their duties, on the whole, there were relatively few casualties. This may have been due to good luck in some cases as, on one occasion, a Canadian firefighting detachment was out fighting fires in the surrounding area when a bomb destroyed their accommodations.

In April 1944, the corps was given the opportunity to send a company of volunteers to augment the Army fire service during the Normandy invasion and ensuing operations on the Continent. A special service company comprising three officers and 98 other ranks was formed and given special training. As events progressed, authorities decided not to use the whole contingent, which was a big disappointment for the volunteers.

Achievements and Recognition

In recognition of the service of this corps, authorization for the issue of two classes of discharge badge was given by order-in-council. One was for those who volunteered and served overseas, the other was for those who volunteered and served in Canada for at least six months, but did not proceed overseas. The

---

16 Major General L.R. LaFleche to S.L. de Carteret, Deputy Minister Air Service, letter dated January 5, 1942.

design for both badges consisted of a silver button in the centre of which appeared an impeller in red enamel, the official insignia of the National Fire Service in the UK. The badge is encircled with the words Canadian Firefighters, 1942–45 and the whole surmounted by a crown. Service overseas was indicated by the addition of a scroll underneath the badge with the word "Overseas."

During the corps' deployment, a total of three members lost their lives and several sustained serious injuries. Commanding Officer G.E. Huff was appointed Officer, Order of the British Empire, and Senior Company Officer N. Torno was appointed Member, Order of the British Empire (OBE). British Empire Medals were awarded to Senior Company Officer M.W. Dolman, Leading Fireman C.J. Diwell and Section Leader J.J. Dewaal. C.J. Diwell was also awarded the Royal Humane Society Testimonial on Parchment for promptitude in saving a life from drowning. Leading Firemen I.F. Cam and W. Bryce were awarded the Royal Society's Testimonial on Vellum for saving a life other than in the course of duty.

Following the departure of the corps, Sir Aylmer Firebrace, Chief of Staff of the National Fire Service, paid the following tribute: "May I say what a pleasure it was to have the Corps of Canadian Firefighters with us. Efficient firemen of fine type and first class physique they have made a thoroughly good impression over here."[17]

As further recognition of the contributions of the corps, the Minister of National Defence authorized the Memorial Cross to include the mothers or widows of members of the Corps of Canadian Firefighters whose death occurred either during their service or subsequent to such service but attributable thereto."[18]

In Remembrance
In September 2003, a group of Canadian Fire Service veterans travelled to London, England, to attend a Service of Remembrance and Dedication of the United Kingdom Firefighters National Memorial at St. Paul's Cathedral in the presence of The Princess Royal. The group joined members of the UK Fire Service in witnessing the addition of three names of the Corps of Canadian Firefighters killed in World War II to the memorial statue. The site of the memorial, south of the Cathedral's South Transept entrance, was chosen for several reasons, including the fact that St. Paul's Cathedral became a symbol and a source of hope and defiance during the Blitz years. The memorial is aptly named Blitz.

The cenotaph memorial depicts two firefighters manning a hoseline with a third calling for more help. It is cast in bronze with a nozzle pointing towards St. Paul's Cathedral. The names of J.S. Coull of Winnipeg, Manitoba, A. Lapierre of Montreal, Quebec, and L.E. Woodhead of Saskatoon, Saskatchewan, have been joined with those of the UK Fire Services who have made the ultimate sacrifice in the course of their occupational duties. The Memorial[19] was unveiled by Queen Elizabeth the Queen Mother on May 4, 1991. The Memorial has been enriched by its elevation and the addition of 1,100 names of firefighters who died in the execution of their duties in peacetime, to those 997 firefighters killed during the last world war, thus creating the United Kingdom Firefighters National Memorial.

---

17 *History of the Corps of Canadian Firefighters*, p. 48. Supplied by the National Archives.
18 Letter, Major General L. R. LaFleche to S. L. de Carteret Deputy Minister Air Service, January 5, 1942.
19 St Paul's Cathedral—A Service of Remembrance & Dedication of the United Kingdom Firefighters National Memorial.

The Blitz monument to fallen firefighters.

The wreath laid at the Blitz monument in 2003.

The Canadian firefighters who attended the ceremony chose Team Mitzi as an identity for the group. Mitzi was a dog that became somewhat of a mascot for the corps and was brought home to Canada after the war. The team members were:

- Jack Coulter, veteran of the Corps of Canadian Firefighters and former chief of the Winnipeg Fire Department;

- Peter K. Ryan, retired district chief of the Ottawa Fire Department and who has been assisting in efforts to establish a Canadian Fallen Firefighters Monument in Ottawa. He has been a member of an advisory committee with the Canadian Fallen Firefighters Foundation, recently formed to begin the long task of registering Canadian fallen firefighters in Canada;

- Alex Forrest, of Winnipeg Local 867 of the International Association of Firefighters, representing 260,000 professional firefighters in Canada and the United States, who marched in the Brigade of Colours outside St. Paul's Cathedral;

- Shannon Pennington, Senior Chief, North American Firefighter Veterans Network (NAFVN) and 26-year career veteran of the Calgary Fire Department, who was team leader and organizer of the Canadian contingent;

- Ms. Alvina Drennen, Fallen Firefighters Association, New York;

- Lieutenant Colonel Lorne MacLean, OMM, CD, (1951–87) who served 36 years in the Fire Service of the Canadian Armed Forces, the final eight years as the CFFM and is a past president of both the Canadian Association of Fire Chiefs (CAFC) and the Canadian Association of Fire Marshals and Fire Commissioners (ACFM&FC); and

- Chaplin Lorne Bjarnason, NAFVN, who served in the Calgary Fire Department for 27 years.

Diana Dakers, retired registered nurse, and Lois MacLean, both from Ottawa, served as team photographers. All members were self-funding.

Belated Action—Canadian Government
The Corps of Canadian Civilian Overseas Firefighters, much like the veterans of Canada's Merchant Navy, have had to suffer delay in recognition by the Canadian government, until the passing of Bill C-41 on October 20, 2000. On November 11, 2003, for the first time, a poppy wreath was laid at the National Cenotaph in Ottawa, in remembrance of the corps. The St. Paul's Cathedral event and the November 11th service in Canada mark a historical step in the process of recognizing the dedication and sacrifices of the Corps of Canadian Firefighters, some 58 years after the end of World War II.

Interment
The three firefighters who lost their lives during World War II—Fireman J. S. Coull, Section Leader A. Lapierre, and Section Leader L. E. Woodhead—are buried in the Brookwood Military Cemetery which is located a short distance outside London. Covering 37 acres, it is the largest war cemetery in the United Kingdom, and is owned and maintained by the Commonwealth War Graves Commission. The cemetery features a Stone of Remembrance at the centre and a Cross of Sacrifice at the top. It is the final resting place for 2,405 Canadians.

At the going down of the sun, we shall say, we have not forgotten them, we shall not forget . . .

The Cross of Sacrifice monument in Brookwood Military Cemetery, flanked by rows of headstones. The inscription: Their Names Shall Liveth for Evermore.

WORLD WAR II

Shortage of Manpower

The battlefields of Europe consumed manpower at an alarming rate during World War II, and the demand for able-bodied men for front-line duty reached epic proportions. Consequently, the luxury of having a full-time force dedicated to firefighting operations had to wait until manpower availability improved and the hazards of fire became a significant threat to operations.

Unprecedented Expansion of the Military

The Second World War heralded a time of unprecedented expansion for all three services. As the Canadian military rapidly progressed to a force of considerable size and power, the fire services also benefited, obtaining increases in manpower, vehicles and equipment. Justification for the rapid expansion came from the frequent aircraft crashes and large fires that occurred during the war. Several of these large fires threatened vital war material stocks. These incidents all served to underscore the Fire Service's

The headstone of Section Leader L.E. Woodhead. The inscription: Not just today but every day in silence we remember.

potential value, demonstrated by limiting the amount of destruction caused by fire.

The Disturbing Fire Threat

Serious fire losses in the early years reinforced the importance of fire protection, especially at places such as Mossbank, Saskatchewan, where the destruction of major buildings occurred. Another equally destructive fire occurred in May 1941 at the air observer school at Portage la Prairie, Manitoba. One week after the completion of Hangar No. 2 and with a dozen new Anson aircraft parked inside, a student pilot flying in high winds suddenly stalled and crashed into the roof. Unfortunately, the dozen twin-engine Anson aircraft had just been refuelled. Within moments, the aircraft gas tanks were exploding, making the hangar unapproachable for meaningful firefighting operations. Prevailing winds managed to spread the fire to a nearby hangar, but through the extraordinary firefighting efforts of the Station Fire Department, augmented by firefighters from nearby RCAF Station MacDonald and the city of Portage la Prairie, Hangar No. 1 and other adjacent buildings were spared. Unfortunately, Hangar No. 2 was gutted. With duly applied wartime haste, the destroyed hangar took just three months to rebuild.[20]

Similar situations occurred across the country, where a number of highly destructive fires added credibility to the continued growth of the military fire service. Construction of swimming pools to provide additional water storage for firefighting needs was among the remedial actions taken after devastating fires at Mossbank and Dafore, Saskatchewan, bombing and gunnery schools. These were the only schools on the Prairies to obtain swimming pools for supplementing firefighting water supplies.[21]

The Canadian public was also beginning to appreciate the importance of a well-equipped, well-trained military fire service. This attitude may have stemmed in part from stories of the London Blitz that were very much in the news. In 1942, the *Yorkton Enterprise* reported, "Fire protection was of considerable importance, particularly during wartime when economy must be the watchword; it is sound logic to invest money in good firefighting equipment rather than suffer the huge losses that are often the result of fire."

The Need for a Firefighter Trade

The need for a distinct military trade to be employed in the Fire Service became clear with the development of large Second World War aircraft and their increased fuel capacity. Naval and army supply depots containing large quantities of munitions and vital war material warehouses also demanded on-site fire protection. Not surprisingly, as aviation technology advanced, so did the need to improve firefighting skills and aircraft crash firefighting equipment.

On the naval and army side of operations, fires in buildings housing essential war material necessitated advances in structural firefighting techniques and improved apparatus performance. Using these new requirements as building blocks, superior proficiency in all aspects of the firefighting service was steadily being achieved, particularly during the 1939–45 war years.

The BRITISH COMMONWEALTH AIR TRAINING PLAN

Overview

At midnight on December 17, 1939, a small group of men gathered in the office of Prime Minister Mackenzie King to sign a document that would launch the British Commonwealth Air Training Plan (BCATP). The primary purpose was to provide an environment for training Commonwealth pilots.

Execution of this plan resulted in a rapid increase in the number of air-training fields and contributed more than 130,000 airmen to the air forces of the allied nations. As the training plan began to take shape, the construction and commission of large numbers of airfields made it imperative to establish a fully organized, well-trained fire service.

---

[20] Peter Conrad, *Training for Victory, the British Commonwealth Air Training Plan in the West* (Saskatoon: Western Producer Prairie Books, 1989), p. 96.
[21] Ibid., p. 98.

There was a feverish bout of activity at this time that reached into every corner of the country, a kind of national single-mindedness that only the threat of total war could generate. Between 1939 and 1944, over 8,000 buildings were constructed, 700 of which were aircraft hangars; 300 miles of water mains were laid; and storage facilities developed for some 26 million gallons of aviation fuel. One of the explanations for the rapid construction of the wartime buildings lies with the superb efficiency of the facility design team in Air Force Headquarters (AFHQ) in Ottawa, which rapidly produced the sets of standard building designs for all BCATP construction across Canada.

Following is a quote by a building contractor for wartime aircraft hangars:

> The hangars came prefabricated from large producers of heavy timber. They were all timber; beautiful timbers in those roofs, lots of fir, four by twelve, four by eighteen, even four by twenty. There was a shortage of steel, they needed that for shells, and so the wood structure fitted in admirably with Canada's natural resource production. That's how the trusses were made, just wood and some bolts.

A construction helper described it this way:

> The hangars came in by rail and then they were trucked out just as you would truck out a load of lumber. They were all shipped pre-cut and all we did was bolt the skeleton together, nail on the siding, and shingle it. The doors were shipped in ready-built. We could put up a hangar in no time.

The final cost of this undertaking of over $2 billion was shared among Britain, Australia, New Zealand and Canada, with Canada paying 72% of the total cost. The spectre of losing newly constructed structures and precious fuel supplies through fire produced the impetus necessary for the unprecedented expansion of the military fire service.

Developing Fire Service Support

Civilian fire departments initially provided professional leaders to run the newly expanded Fire Service. In July 1940, the Ontario Fire Marshal's Office and the Dominion Association of Fire Chiefs sponsored the first of three firefighting courses at the University of Toronto. Once again, the Ontario Fire Marshal played a key role in recruiting and training for the initial intake of nine firefighters.[22] These were all professional firefighters drawn from departments across the country and were given a hastily created three-week course. Recruiting and training continued, and three serials graduated by mid September 1940.

The graduates, mainly firefighters from civilian fire departments, formed the nucleus of supervisors destined to serve in the newly constructed RCAF fire halls.[23] They were awarded the ranks of corporal, sergeant or flight sergeant, depending on the experience or rank they had held in civilian departments. They were then posted to one of the many RCAF stations opening up across the country.

The demand for firefighters for overseas service in the auxiliary fire brigades of Great Britain, plus the continued requirement for firefighters in the municipalities, culminated in a countrywide shortage of trained firefighters. This particular problem became a point of discussion at the Fire Prevention Officers' Conference in April 1941. A key recommendation

---

22 Memoirs of Flight Lieutenant John W. Cowell, CD, "A History of the RCAF Firefighter" (unpublished, 1990).
23 Memoirs of Flight Sergeant C.G. Constable, BEM.

developed during the conference was that the manning would be one non-commissioned officer (NCO), one driver and four airmen for a fire tender while there was one NCO, one driver and three airmen for a fire trailer. However, a memorandum to the Minister[24] suggested the following manning: one flight sergeant, one sergeant, two corporals, five airmen, and two airmen drivers for the fire tenders and one flight sergeant, one sergeant, two corporals, four airmen and two airmen drivers for the fire trailers.

In response to the continuing shortage of trained firefighters, the RCAF opened its own firefighting school in Mountain View, Ontario, a detachment of nearby RCAF Station Trenton. Thereafter, it became common practice to induct firefighter recruits without previous firefighting experience. The frequency of training courses was dictated by the needs of the recently developed scale for manning fire halls. Under those scales, the number of personnel was primarily dependent upon the unit's establishment of mobile fire equipment, rather than keyed to the risk exposure presented by infrastructure and military operations.

The memorandum also included a recommendation that firefighters be given the task of driving the fire vehicle. The existing system employed Motor Transport drivers and did not recognize that the driver had to be a practised pump operator. There was also the potential problem of lack of familiarity with the airfield, since frequently changing drivers made team training extremely difficult. These recommendations were accepted and the firefighters took over the driving duties. Nevertheless, the manning strengths were to continue as a problem for some time. While many stations were adequately manned, others were not so fortunate.

In an effort to upgrade the command structure of the firefighting service, several senior NCOs were commissioned to the rank of pilot officer and appointed as command fire prevention officers (CFPO). The task of these CFPOs was to develop a schedule of inspections for each installation within their jurisdiction and carry out on-site visits. These assessments of equipment suitability and overall department efficiency were then forwarded to AFHQ. It was customary to schedule a meeting of the Station Fire Prevention Committee to coincide with a CFPO's visit. In that manner, the CFPO could act as technical advisor and offer guidance and advice on local initiatives for fire prevention.

The Air Force Fire Service eventually expanded to a maximum strength of 1,700 trained firefighters. By comparison, that was more personnel than the combined total of firefighters employed by the cities of Toronto and Montreal in 1941.[25]

The JOINT FIRE SERVICES FIRE COMMITTEE

Structure
The JSFC replaced the WSFPC in May 1942. One of the main mandates of the JSFC was to standardize the method of applying fire-safety measures on military property. The committee comprised the fire marshals and their representatives from each of the three services. In attendance at the first meeting were: Commander C.A. Thomson, Director of Fire Safety for the Navy; Lieutenant Colonel O.L. Lister, Director of Accommodation and Fire Prevention for the Army; and Squadron Leader J.E. Richie, Works Fire Prevention Officer for the Air Force.

---

[24] Memorandum to the Minister, File: 925-1-72 (DAO) June 1941.
[25] Madeleine Stace, *History of the Construction Engineering Branch*, p. 18.

Role and Philosophy

The brief existence of the WSFPC (1941–42) had revealed a continued need for inter-service co-operation and the sharing of information. The JFSC intended to carry out the initiatives started by the WSFPC, and during the first meeting, the JSFC determined that their responsibilities would be fire prevention and firefighting in keeping with the practice already established in municipal departments. The committee was unanimous in its commitment to man all flying stations with military personnel. Civilian firefighters working for the military services would normally be employed at designated depots in or close to urban centres. Agreement was also reached to continue the practice of using personnel from other trades to help man fire vehicles, as this arrangement had obvious advantages of providing some support for the limited number of firefighters. From the point of view of security and discipline, the committee decided to have the basic cadre of firefighters remain servicemen. These recommendations were re-affirmed in November 1945.[26]

One of the committee's early proposals was to establish a fire safety subcommittee on the east and west coasts. This initiative improved communications and implementation of central committee initiatives. The subcommittees were mandated to meet on a regular basis, to discuss problems concerning fire safety and to make appropriate recommendations. Intractable problems that extended beyond the subcommittees' authority were to be forwarded to the JSFC for further consideration and action. The 1943 Atlantic Subcommittee was made up of the following members: Lieutenant W.J. Carson, Fire Safety Officer for the Navy; Major B. Currie, Fire Prevention Officer for the Army; and Flight Lieutenant T.H. Mathews, Command Fire Prevention Officer for the Air Force. The Pacific Sub-Committee comprised: Lieutenant B.O. Nixon, Fire and Safety Officer for the Navy; Lieutenant Colonel O.L. Lister, Fire Prevention Officer for the Army; and Flight Lieutenant W.A. Carlisle, Command Fire Prevention Officer for the Air Force.

Period of Office

The JSFC and its subcommittees continued their work from inception in 1943 until the three services were integrated in 1966. Their impact on fire safety was profound and the framework for the delivery of fire safety communications was very effective.

INTEGRATION—UNIFICATION—INTEGRATION

Overview

In 1966 the implementation of a radical new government policy, respecting the organizational structure of Canada's military, and which involved the integration of the three armed services, set the stage for a hectic period of activity over the next two years. During this time each service retained its traditional identity, including the uniform worn by its members. This degree of autonomy ended in 1968 when unification came into effect and a uniform that was common to all was selected. Unification remained in effect for two decades until the mid 1980s when the negative effects, such as loss of identity and the attendant loss of esprit de corps, were recognized as outweighing the benefits. Wisely, the logical restructuring that was part of the integration process was retained.

Headquarters Formations

Integration and unification caused some confusion and inter-service rivalry in the newly integrated headquarters as the three services vied for the newly created position of CFFM. Lieutenant Colonel Lindsey-Brown, the former Army Fire Marshal, was named as the first

---

26 Directorate of History, File 5192 15-9-111 (Ottawa, 1945).

CFFM. The organizational structure included: Lieutenant Commander Neil Duval as Deputy CFFM; an Operations section, made up of former RCAF Fire Protection Officers and headed by RCAF Squadron Leader Walter (Wally) Sinclair; Fire Engineering headed by Mr. Arme DeRoche and staffed by civilians; Policy Development, staffed by Lieutenant Commander Alex Hope and Flight Lieutenant Archie Graham; and an Orderly Room for internal administration. Lieutenant Colonel Lindsey-Brown served for a brief term and was succeeded by Navy Commander Neil Duval.

The Regional Concept
During the planning period leading up to integration, Paul Hellyer, Minister of National Defence, sought examples of how integration of the three military services could work. At that time, a study of the construction engineering function that had just been completed, reflected the eye-catching numbers of a reduction of 40% to 50% in manpower inherent in an integrated construction engineering organization.[27] Accordingly, the race to develop specific plans was on in 1964. By March 1965, a detailed plan for regional construction engineering formations had been prepared. By mid 1966, this plan had been implemented, with a Regional Construction Engineering Office (RCEO) at Halifax, Montreal, Toronto, Edmonton and Vancouver.

This was a massive change to the basic organizational structure, from one keyed to the functional divisions of the military formations to one based on geographical considerations. With no commitment to any specific military command or other formation, the offices were more or less constituted as an in-house consulting service. For the Fire Service, being a component of the Military Engineer Branch, this included a change to its organizational structure.

The change required that all fire-related responsibilities, other than for special weapons (nuclear munitions), would be assumed by the RCEOs. There was a Regional Fire Marshal's Office (RFM) at each of these sites. RFMs had responsibilities similar to those traditionally associated with Command Fire Marshals, except that, not being an agent of a military command, they needed to operate in a manner more advisory than authoritative. This arrangement was short-lived and was terminated on April 1, 1970, as the Commands were dissatisfied about the loss of control over these resources. In retrospect, it is not readily evident whether this experiment was driven by an idealistic vision or was the result of a blurring of the pragmatic view of what could realistically be achieved.

Fire Protection Philosophy
The Navy developed a system that operated somewhere between the approaches of the Army and Air Force, concentrating efforts on the shore-based activities directly related to seafaring, while providing protection for dockside infrastructure. The Navy retained this traditional and unique approach respecting fire-protection services.

The Army philosophy tended to vary. Basically, the operation of a fire department was based on the belief that the establishment be directed primarily to firefighting. Fire prevention was not ignored, but inspections and follow-up recommendations were considered a fire-engineering task to be carried out under the auspices of the respective command fire marshals. Therefore, the fire marshal spent many hours coming to terms with the real and perceived problems of each installation. Productive use of the on-site firefighters as fire inspectors was not maximized at that time.

---

[27] *The History of the Canadian Military Engineers*, 1997.

The Air Force policy for fire protection services placed a high degree of importance on fire loss-limiting engineering, fire-safety inspections to enforce applicable regulations and on fire-safety education of the workforce. From an operations perspective, the first priority was placed on rapid response to airfield emergencies. As well, the capability to fight large structural fires was retained and fostered.

As the fire services in each arm of the forces evolved in face of the changing circumstances, the challenge facing the integrated headquarters was to create a philosophy that would provide the greatest benefit to all, in full cognizance of the special requirements specific to each of the three military services.

Equipment Inventories
Combining the three fire services brought together a vast inventory of firefighting equipment, both mobile and static. Each item came with its own unique design as a result of different operational demands. Major pieces of equipment that were familiar to one service were often foreign to another. This included major items such as fire boats, range firefighting vehicles and aircraft rescue and firefighting vehicles (ARFFV). There was considerable commonality respecting structural fire apparatus.

One unfortunate common problem was that much of the existing equipment was worn out or badly in need of replacement. For example, the majority of the pumpers were 20 years old, the fire boats approaching 24 years in operation and inadequate for their designated role, and the major foam vehicles (MFV) were fast approaching the end of their useful life.

There was a pressing need for the unified staff officers to quickly become familiar with the demands of each of the other two services, to avoid misunderstanding.

For example, the RCAF had some time earlier changed the brass nozzles for nozzles made of lighter composite materials. However, these would not stand up in a salt-air environment, so brass had to be retained for naval use. Many similar equipment problems arose in the years immediately following unification. Patience and goodwill were in great demand.

Facts and Emotions
To start with a basic fact, in the main, integration followed by unification only affected the military component of the Fire Service, to the almost total exclusion of the civilian component, which constituted approximately half of the firefighters. For the military firefighters in place at the time, although their role remained as it traditionally had been, many changes, such as the placement of the trade within the lowest paid group of trades, generated painful emotions and caused the morale of the military firefighters to sink to an all-time low. Major J. Torraville, later Lieutenant Colonel and CFFM, carried the brunt of the negotiations to regain the trade's rightful status. It took over two years to re-establish firefighting as a technical trade that was subject to high-risk operations.[28]

Other factors that generated emotional resentment amongst the military firefighters included the melding of two other trades, navy air boatswains and army fire inspectors, into the firefighter trade and the loss of the Air Force blue uniform and the Air Force cap badge. The uniform and badge were replaced with the common-to-all green uniform and the military engineer cap badge. Even so, it was the personnel from these two trades being displaced and moved into the firefighter trade that bore the brunt. Many held quite senior ranks that often had been earned in lesser time than would be the case for Air Force firefighters. This forced many into the unenviable position of being in command of fire department operations and

---

[28] Major Phil Brown (1947–75), *The Royal Canadian Air Force Fire Service 1939–1975: A Subjective History* (unpublished).

supervising personnel with substantially more experience in a profession that they, the newcomers, were unfamiliar with. Both had to learn a new trade and adapt to working in a new environment. It is a great credit to them and to the Fire Service as a whole that so many went on to carve out productive and successful careers.

The RCAF badge and the Military Engineer badge.

Badges that are a part of Fire Service history.

Firefighter Turmoil

To add a perspective based on quantification, at this point, Canada's military services had some 2,200 firefighters. Included in this total were 200 civilian firefighters of the Navy, 500 civilian firefighters of the Army, and 300 civilian firefighters and 1200 military firefighters of the Air Force. It was to this mix that 100 personnel who had previously been navy air boatswains, and 50 personnel who had previously been army fire inspectors, were added.

For the Air Force firefighters, accepting these new tradesmen into their midst from the other services who, quite frequently, were senior in rank with little or no experience in fire department emergency operations, was discouraging and seemingly foolhardy. There was also a degree of coercion brought to bear on firefighters, up to and including those of sergeant rank, to sign a waiver of their conditions of service by agreeing to accept postings to positions aboard naval ships as a means of enhancing their careers or, put another way, to avoid any impediments to their careers.

Collateral Damage

Unification also introduced a sweeping change, not specific to the firefighter trade, which is not often mentioned when the advantages and disadvantages of unification are discussed, yet was and remains of far-reaching importance: specifically, the destruction of the time-honoured rank of corporal. Essentially it became a rating similar to private first class rather than a rank that had long been recognized as the first level of command and supervision. For those who had earned the rank through dedicated service and the merit principle, it constituted an emotional crunch. In the years that followed, those who were given the title more or less automatically were often derogatively referred to by the term "Hellyer Corporal," which took its name from that of the defence minister, Paul Hellyer, who was the force behind the unification experiment. The move was not required in order to implement unification, perhaps it was an ill-concealed form of bribery as a means of buttressing morale in a target group.

With the traditional first level of command and supervision now in tatters, those involved with implementing, what at the time may well have seemed to be mindless policies, recognized that something had to be done. The solution chosen was to create a new rank, which was assigned the title of master corporal. The route ahead was now quite straightforward: simply converting all those who had earned the rank of corporal prior to unification to master corporal. This did not happen. As a result of the hectic circumstances that existed at the time, a stop-gap measure was selected.

The measure authorized individual units to promote selected personnel to master corporal, up to the number of positions previously established for corporal rank. As will be quickly recognized, this adhered to the merit principle at the local level, but not necessarily at the national level. Clearly the stage was now set for yet another moral-busting policy. There would be no disappointment here. When the national promotions list to master corporal was released, it appeared to adhere rigidly to the most recent merit list, seemingly to the near total disregard of what had gone on before. Personnel who previously had held the rank of corporal prior to unification and who were subsequently promoted to master corporal at unit level, and were not on the national promotion list, lost the rank. Essentially, they were busted twice within a few months, in neither case through any fault of their own.

Titles and Insignia of Ranks
The process of converting the Navy, the Army and the Air Force into a single service required, amongst so many other things, that titles and insignia of ranks be homogenized. Overall, this did not have a significant impact, but it did contribute a sense of loss of one's roots and associated feelings of insecurity. A listing of titles of ranks of the pre-unification Navy, Army and Air Force, the unified Canadian Forces, and post-unification navy, army and air force, arranged in a juxtapositional manner, is shown in Figure 1-1.

Bonding
As firefighters of all three military services began to train and work together, the sense of internal rapport and the process of bonding, a recognizable and important feature of the Fire Service which gives it its strength, began to take hold. This was far from a smooth transition, as often egos and prejudices got in the way of mutually desirable co-operation. Nonetheless, within a short time, conditions began to improve and stabilize. Working together, training together and to the extent that is perhaps peculiar to the Fire Service, living and playing together, was a strong foundation for the power of osmosis to unobtrusively eliminate the friction.

Training conducted by the Fire School now included both civilian and military firefighters in a common learning experience at CFB Borden. As time moved along, the students became a mixture of all three services, plus civilians. Irrespective of their background, all students were trained to a common standard. In 1970, the title of the organization was changed to Firefighter Training Company (FFTC)[29] and was designated an integral part of the Canadian Forces School of Aerospace, Ordnance and Engineering (CFSAOE).[30] With the break-up of the CFSAOE in 1985, the FFTC became the CFFA.

Integration and Unification in Review
When unification is viewed in the broadest of terms it is prudent to focus on the positive conditions it yielded rather than on the many rather traumatic policies that were imposed on the three military services along the way. Although there was a common military uniform, it

---

[29] Major Phil Brown (1947–75), *The Royal Canadian Air Force Fire Service 1939–75: A Subjective History* (unpublished).
[30] With the break-up of the CFSAOE organization in 1985, the FFTC became the CFFA.

Figure 1 - 1
RATINGS & RANTS ■ COMPARATIVE TITLES

Pre-Unification			Unified	Post Unification		
Navy	**Army**	**Air Force**	**Canadian* Forces**	**Navy**	**Army**	**Air Force**
Ordinary Seaman	Private	Aircraftsman 2nd Class	Private	Ordinary Seaman	Private	Private
Able Seaman	Private	Aircraftman 1st Class	Private	Able Seaman	Private	Private
Able Seaman	Private	Leading Aircraftsman	Corporal	Leading Seaman	Corporal	Corporal
Leading Seaman	Corporal	Corporal	Corporal	Leading Seaman	Corporal	Corporal
			Master Corporal	Master Seaman	Master Corporal	Master Corporal
Petty Officer 2nd Class	Sergeant	Sergeant	Sergeant	Petty Officer 2nd Class	Sergeant	Sergeant
Petty Officer 1st Class	Staff Sergeant	Flight Sergeant	Warrant Officer	Petty Officer 1st Class	Warrant Officer	Warrant Officer
Chief Petty Officer 2nd Class	Warrant Officer 2nd Class	Warrant Officer 2nd Class	Master Warrant Officer	Chief Petty Officer 2nd Class	Master Warrant Officer	Master Warrant Officer
Chief Petty Officer 1st Class	Warrant Officer 1st Class	Warrant Officer 1st Class	Chief Warrant Officer	Chief Petty Officer 1st Class	Chief Warrant Officer	Chief Warrant Officer
Sub- Lieutenant	Lieutenant	Flying Officer	Lieutenant	Sub- Lieutenant	Lieutenant	Lieutenant
Lieutenant	Captain	Flight Lieutenant	Captain	Lieutenant	Captain	Captain
Lieutenant Commander	Major	Squadron Leader	Major	Lieutenant Commander	Major	Major
Commander	Lieutenant Colonel	Wing Commander	Lieutenant Colonel	Commander	Lieutenant Colonel	Lieutenant Colonel
Captain	Colonel	Group Captain	Colonel	Captain	Colonel	Colonel
Commodore	Brigadier General	Air Commodore	Brigadier General	Commodore	Brigadier General	Brigadier General
Rear Admiral	Major General	Air Vice Marshal	Major General	Rear Admiral	Major General	Major General
Vice Admiral	Lieutenant General	Air Marshal	Lieutenant General	Vice Admiral	Lieutenant General	Lieutenant General
Admiral	General	Chief Air Marshal	General	Admiral	General	General

* The navy continued to use its traditional titles in the workplace and individuals could retain the navy title specific to their rank, providing a formal request received approval.

did not change the roles of the navy, army or air force. How could it be otherwise? Perhaps the only change of substance occurred with the implementation of integration and remained through unification and back to integration was the formation of an NDHQ. To narrow our perspective to just the Fire Service, the office of the CFFM was created, and a common path of professional development, now manifested in the CFFA, was implemented. Beyond that, the fire services of each of the military services remained, and remain, virtually unchanged.

Unification did not make any obvious changes to the fire departments. The fire service of the navy, its culture, practices, procedures and idiosyncrasies, underwent little, if any, change, while retaining its own fire marshal. More or less the same can be said for the army and air force. The air force was seven years into unification before re-establishing AFHQ and morphing it into Air Command. The navy and army, virtually from the outset, simply renamed their headquarters as Maritime Command and Mobile Command, respectively.

Unification did not measurably increase interchangeability amongst the members of the Fire Service. Were it not for the air force, which continued to administer its fire service as it had done prior to unification by sending military firefighters to wherever the need was greatest, which now at times included assignments aboard naval ships, it could be argued that there was little, if any, interchangeability of firefighters amongst the three military services which the term "unification" would imply was the case.

The STAFFING CRISIS

How it Developed

A shortage of commissioned Fire Service officers (FSO), particularly at the senior officer levels, precipitated a staffing crisis that first became evident in 1995. At that time, a construction engineering officer (CEO) was appointed to the position of CFFM. Four years later, another CEO was appointed commandant of the CFFA. As is shown in Appendix A, this has continued unchanged. All subsequent appointees to these positions have been CEOs. There were three main factors that led to this situation. Specifically:

- the untimely exodus of FSOs to more attractive jobs in the private sector, a situation exacerbated by the University Training Plan Men program (UTPM);
- lack of sufficient successful candidates for the Commissioning from the Ranks Plan (CFR);
- the application of the Special Ranks Commissioning Plan (SRCP); and
- the loss of FSO-established positions that took place during the restructuring and reductions to the military that was most evident during the mid 1990s. This resulted in the Fire Service Officer Corps being decimated, losing one half of the positions established for the rank of major and one third of the positions established for the rank of lieutenant or captain. This is reflected in Figures 1-2, 1-3, 1-4 and 1-5.

The Effects

Taken singularly, any of these factors has the potential to create staffing problems. Collectively, they have had disastrous results. The first factor, although it had a significant impact, could well have been a phenomenon that would not be sustained and would therefore have self-corrected over time. In this regard, the much valued UTPM is a double-edged sword, which has had a significant influence on the exodus. Graduates have been much sought after by the private sector. Typical of this, during the nine-year period between 1990 and 1999, six FSOs commissioned under this plan left the Fire Service and collectively took a potential 85 years of service with them.

The SRCP is difficult to disparage, since it promotes performers of proven quality into key positions. Regrettably, although unwritten and perhaps ill-recognized, it contains an insidious poison pill in that each established position staffed through the SRCP is one less officer with the potential of progressing through the ranks to the more senior FSO appointments. For a large officer corps, the SRCP could be applied with virtual impunity; however, with an officer corps of such limited size as that of the Fire Service, its application must be done with extreme caution and be very limited if decimation of that corps is to be avoided.

The mathematics of the third factor is such as to speak for itself. As an aside to this, in the early 1990s, a position established for the rank of major in the office of the CFFM was converted to a civilian position. Shortly thereafter, the position was staffed by the person who had previously served in the position as a major. This person retired in 2003. Whether or not subsequent appointees to this position will have qualifications comparable to an FSO, with the experience and fire-service knowledge associated with the rank of major, is problematic. Consequently, the potential for further loss of expertise is a matter of concern.

Figure 1 - 2
Exodus of Fire Service Officers—Mid 1980s—2003

Name	Plan	Rank	Years in Rank	Year Resigned	Years to CRA[1]
Gord Gazley	CFR	Captain	10	1986	10
Don Carmichael	CFR	Major	4	1986	3
Leo Arsenault	CFR	Captain	11	1988	2
Marcel Ethier	CFR	Major	6	1988	11
Dave Geddes	CFR	Captain	9	1989	7
Jack Henderson	CFR	Major	6	1989	7
Chris Halliday	UTPM	Captain	2	1990	20
Jim Wright	CFR	Major	1	1990	8
Herb Livingston	SRCP	Captain	1	1990	1
Bob Giguere	CFR	Major	6	1991	10
Charlie McNeil	CFR	Major	11	1993	3
Andy Beaudin	CFR	Major	2	1993	3
Hal Singleton	CFR	Lcol	7	1993	5
Ev Evans	CFR	Captain	15	1993	0
Jim Johnson	CFR	Captain	13	1993	8
Garry Mauch	UTPM	Major	6	1994	7
Gaetan Perron	CFR	A/LCol	2	1995	2
John Gordon	CFR	Major	7	1995	6
Dick Sadler	CFR	Major	5	1995	8
Terry Gray	CFR	Captain	6	1996	3
Bob Pineault	CFR	Captain	10	1996	4
Yvon Auger	SRCP	Captain	7	1996	3
Ed Larocque	UTPM	Captain	3	1996	22[2]
Denis Girouard	CFR	A/Major	3	1997	13
Barry Colledge	UTPM	Major	2	1999	3
Mark Regimbald	UTPM	Captain	5	1999	18
Neil Drachenberg	UTPM	Captain	6	2000	15
				Years of Service Lost	**221**

[1] Compulsory Release Age
[2] Not confirmed

**Figure 1 - 3
Fire Service Officers—2003**

Name	Plan	Rank	Years in Rank	Years to CRA[1]
Gaetan Morinville	UTPMCM	Major	2	14
Mike Blow	CFR	Captain	6	4
Gary Oliver	SRCP	Captain	2	4
Al Rau	CFR	Captain	6	11
Ken Hoffer	SRCP	Captain	4	3
Don McNeil	CFR	Captain	10	9
Wayne Paterson	CFR	Captain	3	13
Claude Thibeault	CFR	Captain	1	14
Steve Vollhoffer	UTPMCM	Captain	4	12
Denis Pleau	CFR	Captain	2	
Tony Thornhill	CFR	Captain	10	9
M. Mongeau	CFR	Captain	5	13
J. G. Grondines	CFR	Captain	4	13

[1.] Compulsory Release Age

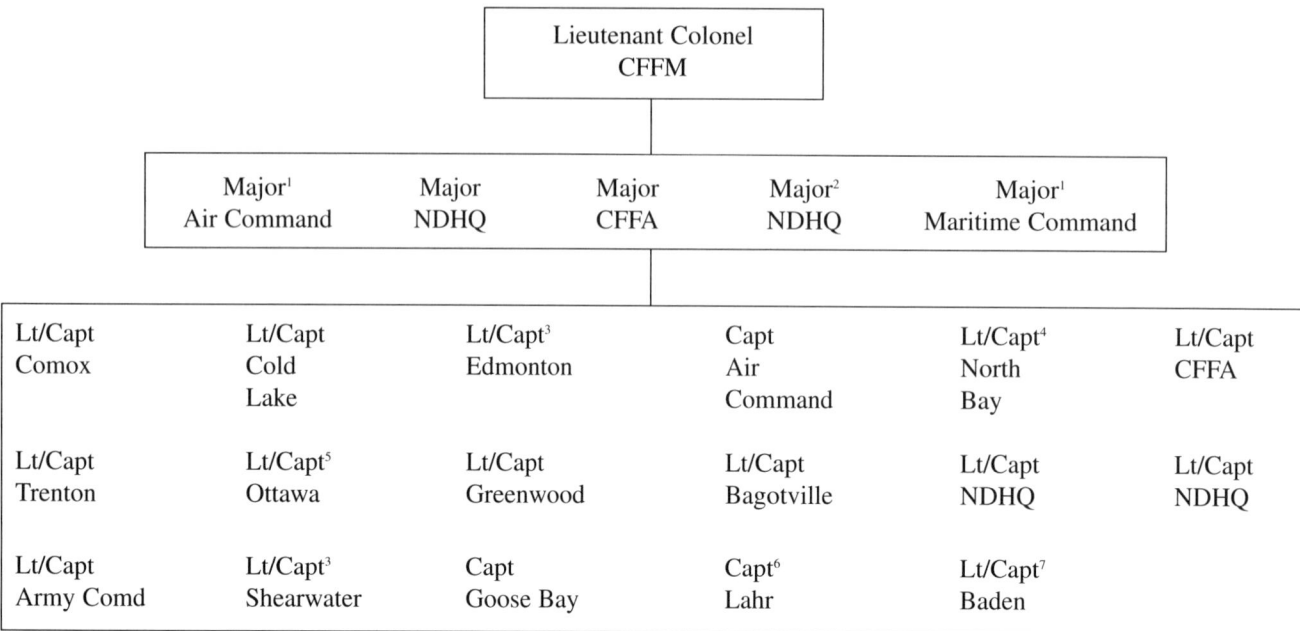

**Figure 1-4
ESTABLISHMENT PYRAMID—Mid 1980s**

1. Deleted during restructuring, coincident with the downsizing and restructuring of the Military Commands in the mid 1990s.
2. Converted to a civilian position in early 1992. Two years later, the fire engineer position was converted from civilian to major. The net result was no change.
3. Deleted, coincident with a reduction in air operations in the mid 1990s.
4. Deleted, coincident with a reduction in and finally cessation of air operations in 1994.
5. Base closure in 1995.
6. Closure of Canadian Forces Europe (CFE) in 1994.
7. Establishment changed to CWO for a brief period. Base closed in 1993.

Figure 1-5
ESTABLISHMENT PYRAMID—2003

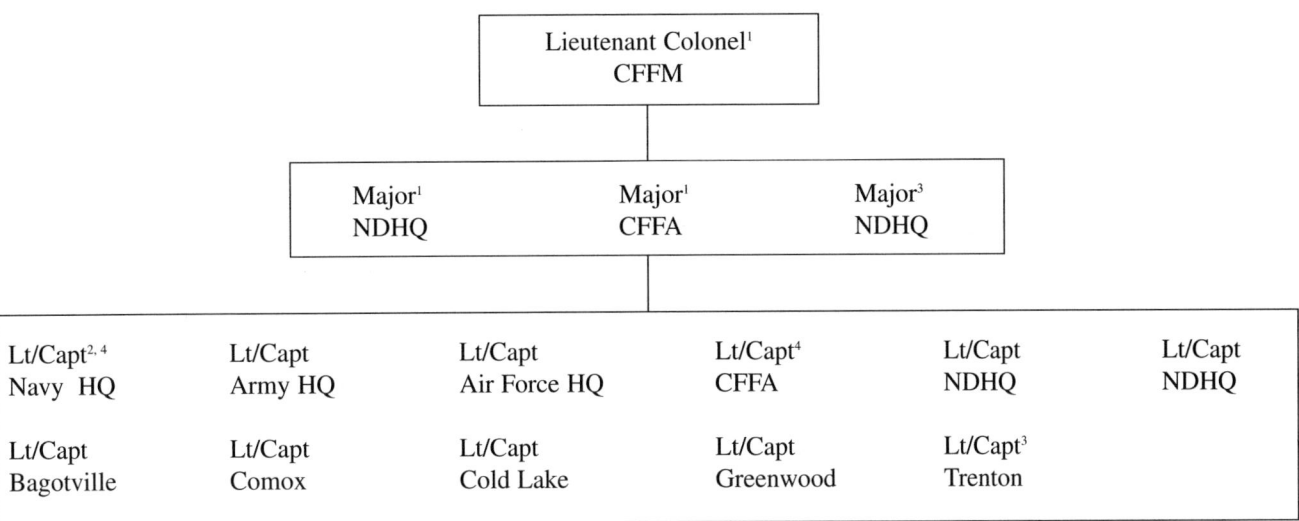

1. Staffed with a CEO.
2. Established during restructuring coincident with the downsizing and restructuring of the military commands in the mid 1990s.
3. UTPM.
4. SRCP.

CHAPTER II

NAVY FIRE SERVICE

WORLD WAR I

Firefighting Force
During the era of the First World War, Navy firefighting responsibility was in the hands of the Engineering Branch (Stokers), supported by damage-control personnel. This applied whether onboard His Majesty's Canadian Ships (HMCS) at sea or in harbour. Essentially, the safety of the ships rested with the officer-of-the-watch supported by duty-watch personnel. Duty-watch personnel were supplied by various divisions within the ship's company.

Naval shore establishments, other than HMC Dockyards Esquimalt and Halifax, functioned as ships ashore in operation and designation bearing the HMCS preface, such as HMCS *Stadacona* and HMCS *Naden*. Similar to the ships, the land-based units also maintained an officer-of-the-day (watch) supported by duty-watch personnel. However, because the workforce at the two main dockyards was mainly civilian, the duty-watch personnel came from docked naval ships. Senior management at the dockyard invariably consisted of RCN officers, and the practice of appointing an officer-of-the-day was universally established.[1]

Protection by Municipalities
While the Navy shore establishments in Halifax were within the area protected by the City of Halifax Fire Department, similar facilities in Esquimalt did not fall within the boundaries of a municipality protected by a

---

[1] Memoirs of Lieutenant Commander Gordon Lay, May 1993.

professional fire department. As a result, a monetary arrangement was formulated with the City of Victoria to provide firefighting services, even though this meant a response run of some eight to ten miles.

Although the arrangement made with the City of Victoria was far from the desired solution, it did work out for the period in question. The two main reasons for the success of these operations were the naval occupants were very fire-conscious, due to their strict sea training, and because the shore establishments were constructed largely of non-combustible brick and were well separated from each other.

Most of the naval firefighting services were almost exclusively directed toward fire on board ships. Firefighting activities when a ship was dockside were treated as a shipboard fire with the same kind of crew response that would occur at sea.

BETWEEN WORLD WARS I AND II

Limited Resources
The Fire Service of the Navy weathered the period between the wars with limited resources. At sea, responsibility for suppression of a fire fell to the ship's Damage Control Team, and it was the practice for the entire ship's company to turn out to aid firefighting operations. This was an understandable approach, given the limited options open to the crew should a fire get out of control. Firefighting operations when the ship was docked were generally assigned to the duty watch of the individual ship or the dockside fire-response crew. The crew members of sister ships in port would be recruited to lend a hand, at least until the dockside or local municipal fire department arrived.

WORLD WAR II

Organizing and Evolving
Prior to the beginning of World War II, the Navy's dockside firefighting was the responsibility of the Royal Canadian Engineers (RCE). This situation was an outcome of the RCE being responsible for construction and maintenance of all shore billets. Captain P.C. Ahern was Advisor for Fire Safety, a position quite similar to that of a modern fire marshal. In 1939, the Navy's dockside firefighting capability still rested with the shore billet volunteer service, duty-watch and the local municipal fire department. It was not until 1942 that a land-based Naval Fire Service began to take shape and eventually accepted complete responsibility for dockside fire protection.

The Naval Fire Protection Services consisted of duty watch (fire piquet) and a designated driver/operator as a "first response unit." Indifferent attitudes with respect to fire-safety measures, however, prompted the Defence Secretary to invite the Dominion Fire Commissioner to carry out fire inspections of the Halifax and Esquimalt dockyards and to submit a detailed report with appropriate recommendations. Coincidently, the WSFPC was formed around the same time and eventually assumed this task. The WSFPC committee had a positive impact on all fire-related matters in the early war years, shaping future policies for all three arms of the military fire service.

By early 1941, naval activities on the West Coast had increased to the point where authorities were openly expressing their concerns over the lack of fire protection for the dockside facilities. These concerns were based largely on fire inspection reports submitted by contracted fire inspection agencies. For their part, senior naval personnel were actively promoting the creation of a dockyard fire brigade that would consist entirely of naval personnel. In 1941, Naval

Headquarters approved the hiring of civilian firefighters to act as fire chiefs at HMC Dockyards in Esquimalt and Halifax.

Laying the Foundation
It took significantly longer for the Navy to appoint a naval fire protection expert to head its firefighting operations than it did for the Army and Air Force. In 1941, both the Army and Air Force had decided to employ a specialist as fire marshal at each of their respective headquarters. However, it was 1943 before the Navy created a similar position. Naval doctrine continued to be focused on the realities of operations at sea.

By 1943, Naval Headquarters recognized the need for a more effective and better organized fire service. As a first step, a position titled Fire Prevention Engineer (the equivalent of fire marshal) was established within the Directorate of Naval Organization. Initially it was filled by Mr. Christian Aldron Thomson from the Office of the Ontario Fire Marshal. He was assigned the rank of lieutenant commander and posted to HMCS *Bytown* (Ottawa), effective May 12, 1943.

A short time later, a Directorate of Fire Safety was established at Naval Headquarters headed by Acting Commander C.A. Thomson Royal Canadian Navy Volunteer Reserve (RCNVR). Mr. John Dawson Crowthers, a lieutenant with the City of Victoria Fire Department, was hired as the Dockyard fire chief at HMC Esquimalt Dockyard. Later, in 1943, Mr. Edward Beals of the Toronto Fire Department was hired to fill the fire chief's position at Halifax. Both were accepted into the RCNVR with the rank of lieutenant, thereby heading the Fire Services of Pacific Command and Atlantic Command, respectively.

Firefighter Recruiting
When the time came to hire additional firefighters, Naval Headquarters policy directed that the dockyard fire department at Esquimalt could only hire men of "Grade 4, fit for duties of light or sedentary nature" (4F). This policy reflected the chronic manpower shortages prevalent in Canada during the war years, with all three services and every sector of private industry vying for available manpower. In many cases, these men were also unfit for service as firefighters. As well, the civilian firefighters could quit their jobs at any time, complicating an already difficult situation. The obvious dangers and inherent problems associated with this situation finally became apparent to Naval Headquarters. In the meantime, the fire department members soldiered on.

At the outset of 1942, Stoker Gordon Lay, RCNVR, and Stoker Norman Stewardson, RCNVR, both professional firefighters from the Oak Bay, British Columbia, Fire Department, were posted to HMCS *Givenchy*, a barracks for HMC Esquimalt Dockyard. Both men were promptly promoted to leading stokers and assigned to Mr. Crowther in the capacity of deputy fire chiefs. One of their first tasks was to divide the 4F civilian recruits into two shifts to operate on a 24 hours on duty, 24 hours off duty schedule. They were also tasked to provide an intense training program using the very limited equipment that was available to them.

The mobile fire apparatus consisted of two, two-wheeled wooden carts and a couple of hose reels. One wooden cart carried a 45 ft / 14 m two-section wooden pole ladder, a 24 ft / 7 m wooden extension ladder, and an 8 ft / 2.5 m roof ladder. The other cart carried a miscellaneous assortment of fire equipment, which included nozzles, axes, wrenches, reducers, and so on. On receipt of a fire alarm, these hand-mobile pieces had to be towed or pushed by the crew to the fire scene. To enable the fire crews to effectively fight ship fires or fires involving jetties, there were two, 250 G/min / 1140 L/min, two-wheeled, trailer pumps housed inside sheds near the jetties. Due to the hilly terrain surrounding the dockyard, they each required at least four men to manoeuvre them around the immediate area. It is of interest to note that hose-reel carts were still in use at some Navy stations and ammunition depots until the mid 1980s.

Conditions of Employment

The duty fire crew quarters at the HMC Esquimalt Dockyard were located on the site of an old residential district that included a few small stores. These buildings covered an area of some four to five city blocks that had been expropriated to allow for the anticipated dockyard expansion. The buildings were demolished with the exception of two small one-storey bungalows. One was used for fire officer accommodation, stores and a training room, the other for crew quarters and an eating area. These very old structures had been built over a three-foot crawl space that was very dirty and contained years of accumulated rubbish. Many hours of hose practice were spent by the duty crew flushing out the huge rats that lived there. The rodents had apparently taken a liking to living in close proximity to the resident firefighters.[2]

With poor living quarters, makeshift equipment and physically limited crew members, it became a trying time for the three professional fire officers. Things could not have been much worse, particularly when contemplating the possible outcome of a serious fire. Notwithstanding all the negatives of the situation, the 4F crews deserved full credit for their obvious enthusiasm and willingness to succeed at a job they were not physically able to perform to the fullest extent.

Firefighter Requirements Reassessed

In addition, the need for personnel who were fully able bodied and provided with modern firefighting equipment was recognized. The 4F personnel were released and replaced by RCNVR enlisted men. Many of these men came from professional fire departments located in towns and cities across Canada and Newfoundland. The newly appointed RCNVR firefighters were given a special designation FF, meaning they could only be posted or drafted from one position to another by authority of the Director of Fire Safety. The FF designator effectively prevented the individual from being diverted to other areas of employment.

Expansion and Training

The Navy, like the other two services, experienced rapid expansion. New bases and other facilities were opened and their fire departments were staffed by Navy Fire Service personnel. At the same time, the departments were taking delivery of new structural fire apparatus manufactured by American-LaFrance and Bickle-Seagrave. The Navy Fire Service was coming of age with all bases and facilities now protected by trained firefighters who were housed in well-equipped fire stations. Supervisors, for the most part, were professional firefighters, thus the level of dockyard fire protection had improved substantially.

Coincident with the expansion came the need to man many new vessels. The requirement to place these ships on active duty as quickly as possible meant that the crews were often required to finish their firefighting training while at sea on operational missions. This state of affairs caused a great deal of concern, especially with respect to fire at sea, but was not fully addressed until 1944.

By 1944, the US Navy had established Class 1 Damage Control Schools to train naval personnel to cope with shipboard damage caused by fire. The training covered a whole gamut of disaster at sea, including damage by fire, explosion, ramming by another ship, shell-fire, and torpedoing. Emphasis was placed on doing the actual

---

[2] Memoirs of Lieutenant Commander Gordon Lay, May 1993.

task. Chief Petty Officer Lay (FF) attended one of these US damage-control schools at the Naval Dockyard in Bremerton, Washington. On his return to Esquimalt, he organized the first Navy fire school that offered short firefighting training courses to ships' crews.

The Esquimalt Fire School
Initially, the school was located on a narrow, gravelled spit of land and was extremely rudimentary. It was housed in a wooden shed containing a trailer-mounted 250 G/min / 1140 L/min gasoline-driven pump. Nearby was a four-foot high circular metal tank about 18 ft / 5.5 m across, half-filled with water and topped off with a couple of inches of diesel oil for training purposes. When the diesel oil was lit, the trainees were led to the fire equipped with a 2½ inch / 65 mm hose, and a Griswald fog nozzle, the standard ship's firefighting equipment. Each trainee crew member had to fight the fire first with the wind at his back and then, after the tank was re-lit, they repeated the procedure, but advancing into the wind. Brief instruction on the Rockford foam nozzle and CO_2 fire extinguishers was also included in the training.

The training was simple, but it served the purpose, as it exposed the seamen to realistic types of situations they might encounter at sea. The main thrust of the training was to prove to the trainees that, properly applied, the equipment they would find on board the ships could be used effectively.

Personnel serving on the East Coast attended a similar US Navy damage-control school in Philadelphia for their initial instruction. More fortunate than their West Coast colleagues, they were able to obtain the use of old ships to carry out more realistic training.

After the war, the RCN opened permanent and well-equipped damage-control schools, similar to their US counterparts, and located them on the East and West Coasts. The training structures included large steel tanks and sections of ships or steel structures made to resemble ships. When the schools opened they were both headed by a naval fire service officer. Unfortunately, post-war changes resulted in these positions being changed from a specialist officer to a general-service officer. Nevertheless, the training continued into the 21st century, albeit somewhat refined while basically following the original framework.

Navy Fire Stations
In 1939 the RCN had, outside Naval Service Headquarters in Ottawa, only two major areas of operations. These were HMC Halifax Dockyard and HMCS *Stadacona* in the Halifax area, and HMC Dockyard and HMCS *Naden* in the Esquimalt area. The dockyards performed maintenance, repair and supply services, while HMCS *Stadacona* and HMCS *Naden* served as training, administration and manning pools in support of the fleet. As the war progressed many other strategic installations opened to better serve the expanding fleet. The Navy deemed it desirable to have Naval Fire Departments protect the facilities at each of the following installations:

West Coast
HMC Dockyard Esquimalt, British Columbia
HMCS *Naden*, British Columbia
Prince Rupert Base, Prince Rupert, British Columbia

East Coast
HMC Dockyard, Halifax, Nova Scotia
HMCS *Stadacona*, Halifax, Nova Scotia
HMCS *Cornwallis*, Cornwallis, Nova Scotia
HMCS *Avalon*, St. John's, Newfoundland
HMC Dockyard, St. John's, Newfoundland
Naval Ammunition Depot, Dartmouth, Nova Scotia
RCN Ammunition Depot, Bedford, Nova Scotia
Naval Ship Repair Base, Shelburne, Nova Scotia
Naval Ship Repair Base, Point Edward, Nova Scotia

Naval Armament Depot, St. John's, Newfoundland
Two RCN Hospitals at St. John's, Newfoundland

Quebec
Signals and Training at St-Hyacinthe, Quebec

This list of installations gives some idea of the enormous expansion of the naval facilities early in the Second World War. This required a corresponding expansion of the Fire Service to protect them and the manpower expansion from a nucleus of 4F civilians in 1942 to the Naval Fire Service status in 1944 was nothing short of extraordinary. The Navy was very fortunate to have been able to recruit heavily from the ranks of the civilian departments in towns and cities across the country.

Waterborne Units
In pre-war years, reliance for fire protection was placed on a shore-based, firefighting capability equipped to draft large quantities of sea water to combat fires in and around the dockside area. During the Second World War, the National Harbour Board maintained and staffed a boat in Halifax Harbour to protect shipping and dockside properties. It was an important role, as the large numbers of ships gathered in Halifax Harbour in preparation for crossing the Atlantic in convoys and the immense quantities of vital war material stockpiled at dockside made the presence of the fire boat an indispensable part of the operation.

Aggressive Shopping
Shortly after the attack on Pearl Harbor in December 1941, all Japanese fishing boats on the west coast of Canada were seized and their crews sent to internment camps. One fairly new vessel that was impounded was an ocean-going, 70 ft / 21 m fish packer with a bridge-mounted, firefighting deck-monitor. This vessel was commandeered, renamed HMCS *Universe*, and became part of the Navy's Fishermens' Reserve, operating out of Esquimalt Harbour. She was never used, however, in a firefighting capacity.

Harbour Traffic and the Japanese Threat
As on the East Coast, the West Coast experienced an enormous increase in shipping and naval activities and saw tremendous increases to the port and harbour facilities. This, and the perceived ever-present threat of a Japanese naval attack, made the need for waterborne firefighting capability essential. Before 1943, most of the dockside firefighting capacity was provided by drafting water with the LaFrance pumpers, but the limitations placed on this operation through low-tide water levels forced the authorities to search for alternate methods. In 1943, a large wooden barge with an immense deckhouse was obtained. The deckhouse was used to cover two 1,000 G/min / 4500 L/min centrifugal fire pumps, driven by a V12 gasoline engine. When the vessel was engaged in firefighting operations, protection was afforded the deckhouse by a water curtain. There were two monitors permanently mounted on an elevated platform and eight 2½ inch / 65 mm hose outlets on the deck. Although it stayed in service until the early 1950s, it had two serious defects. First, it had to be towed by a tug to the area of operation. Second, unless it could be secured to a dock, jetty or other fixed object when the monitors were in operation, it was almost impossible to keep it from drifting off station because of the enormous force exerted on the barge by the reaction cause by the discharge of the monitors.

POST WORLD WAR II

Restructuring[3]
In the spring of 1946, the naval personnel staffing HMC Halifax Dockyard Fire Department were being released from active service. (There were similar

---

[3] Archives of HMC Halifax Dockyard Fire Department.

situations at HMCS *Stadacona*, Bedford Magazine, Naval Ammunition Depot, and in Navy establishments at Cornwallis, Shelburne, St. John's and Sydney.) In consideration of this, a decision was reached to man the fire department with civilians. Accordingly, a recruiting drive began through the Civil Service Commission, with the first employee, William Patrick Delaney, being hired in April. Recruiting was completed in May, with some 200 men being hired. All of those hired were recent members of the Navy, Army or Air Force, as preference was given to those who had overseas service, most of whom needed jobs.

The naval officer in charge at the time was Lieutenant Commander William Simphins, and his assistant was Warrant Officer Redden. These two men screened the applicants and did the hiring. The starting salary seemed fairly decent at the time. It was the huge sum of $1,444 per annum plus a $200 cost-of-living allowance. The work week was based on 24 hours on duty and 24 hours off duty for an average 84 hours per week.

The first officers hired were Lionel Parker, a petty officer in the Naval Fire Department, followed by Nelson McNeil, from the City of Halifax Fire Department, and Douglas K. Lockyer, an ex-RCAF firefighter. These three lieutenants, along with Warrant Officer Redden and the still naval members of the department, conducted a continuous and vigorous training program. This was carried out daily, and the recruits worked from 0800 hours to 1800 hours. At that point, the recruits were sleeping at home as the fire hall still quartered the naval personnel.

The civilian force took over the fire hall sometime in May, and on the very first day, the ultimate embarrassment occurred: the entire attic of the fire hall caught fire and burned. Personnel in the Administration Building saw smoke coming from the eaves and cupola, but said to each other, "the firemen are training again." This fire was extinguished hastily by the new firefighters, once its existence was known.

The Naval Fire Hall HMC Halifax Dockyard. The clock in the tower atop the fire hall, which was built in 1943, has a much earlier birth date than this mid 1950s photo—1767 to be precise. As a result of the planned demolition of the fire hall in 1986, an article published in the Halifax Herald under the headline, "Canada's oldest clock will get a new home," included some interesting details. It reported that its pendulum was nine feet long, its face had a four-foot diameter, it weighed 250 pounds, and had had some 11,000 weekly windings.

The *Montreal Star* headline read, "Firemen see red as Fire Hall Burns."

The actual cause of this fire was never discovered, and there were persistent rumours that it was a cover-up for inventory shortages. At the time, all fire department stores were in the attic. Regardless of the cause, the effect was appalling, and it took many years of hard and serious work before this incident was forgotten and ships' officers began to regard the Naval Fire Service as an effective firefighting unit.

The equipment in service at the time consisted in part of the following apparatus:

- two 750 G/min / 3400 L/min Bickle-Seagrave pumpers
- one 1,000 G/min / 4500L/min American-LaFrance pumper
- one 65 ft / 20 m American-LaFrance Aerial Ladder truck
- one Bickle-Seagrave quad
- one wooden hulled fire boat, *The Nashwack*
- several trailer pumps of various makes
- one one-ton pickup
- one jeep
- one staff car

The engine on most of the apparatus had 12 cylinders, twin-ignition, dual fuel pumps and a dual electrical system, designed to ensure the vehicle would start and make it to a fire scene. However, all had open cabs and were miserable to drive in or on when the weather was cold or wet. All apparatus was built like this at the time.

While the Naval Fire Service was undergoing the changes mentioned, Chief Churchill of the City of Halifax Fire Department retired. This led to a contest between Fred C. MacGillvary and Joseph W. Harber for Chief of the Halifax Fire Department. MacGillvary got the appointment and, as a result, Joseph W. Harber was hired by the Department of National Defence (DND) with rank of lieutenant commander to head up the Naval Fire Department. Harry Curran, another City of Halifax firefighter, was hired as a fire captain to aid in the building of the Naval Fire Service. Both Chief J. Harber and Captain H. Curran served overseas during World War II as volunteer firefighters during the great Blitz of London.

With the hiring of these experienced men, the formation of the civilian-manned fire department was complete, and the last naval member, Lieutenant Commander W. Simphus, went to NDHQ in Ottawa as head of the Naval Fire Service. At the time, the Navy, Army and Air Force each maintained a separate fire service.

The average age of the new firefighters was in the 26-year to 27-year-old range, and most were single. Having all recently been on active service in the Navy and accustomed to dealing with the stresses of armed conflict as best they could, they were mostly still a pretty wild and hard-drinking group who had not yet settled back to civilian life. Many disciplinary actions resulted, but many more infractions were not discovered. However, as time passed, most of those remaining married and settled down to make a career of firefighting.

One of the major jobs facing the restructured Fire Service was to complete an initial fire inspection report on all existing buildings. These reports required much detail which included the size and type of construction, number of stories, heating, interior design and details of any fire protection systems. This established an information bank for all buildings and the same type of report is still completed when a new building is constructed. An ongoing training program was also established, and like the inspection program, is still carried out in the departments.

The restructured Fire Service was made responsible to the Queen's Harbour Master (QHM) and remained as such till integration in 1967. Disciplinary charges were heard by the QHM and any punishment was meted out by him.

Changes to the Waterborne Component
At war's end, the Navy had an outstanding contract with an Ontario shipyard to build three harbour tug vessels the Navy no longer required. This was somewhat fortuitous for the fire departments, as the

three steel-hulled ships were then modified to make excellent fire boats with the addition of fire pumps, deck monitors and fire-hose connections. The dockyards at Esquimalt, Halifax and St. John's each received one of these vessels. Later, 2½ inch / 65 mm and 1½ inch / 38 mm hose reels and fire equipment lockers were installed.

These ships gave much needed service to the Navy and civil authorities during their tour of operation. The Esquimalt-based boat was dispatched on two occasions to the City of Victoria waterfront to combat fires that had progressed beyond the control of the city's shore-based apparatus. Similarly, the Halifax fire boat fought many fires in the harbour area. So important was the fire boat to the City of Halifax that it paid DND $40,000 per year to secure its services until the more dangerous wooden piers and most of the highly combustible buildings had been removed.

The naval fire boat in St. John's harbour was disposed of in 1962; the vessels at Esquimalt and Halifax were replaced in 1975 with two fire boats built in Vancouver. The Fire Marshal of Maritime Command, Gordon Lay, was assigned the responsibility of consolidating all the firefighting requirements on the two vessels. These new boats incorporated a vastly improved firefighting capability, with a 5,000 G/min / 22,275 L/min pumping capacity, delivered through three deck-mounted monitors, or twelve 2½ inch / 65 mm hose connections mounted on the vessel's deck house.

Mr. Gordon Lay.

Bill Fitzpatrick Memorial
The fire hall in HMC Halifax Dockyard is named the Bill Fitzpatrick Building after a former fire chief. FR7 Bill Fitzpatrick (1946–89) joined the Fire Service at HMC Halifax Dockyard in 1946 and rose through the ranks to become fire chief (1976–89). He served for 43 years with the Fire Service and passed away in November 1990. Dedicated on May 19, 1994, it is believed that this is the first DND fire hall to be named after a fire chief.

The Naval Fire Hall, HMC Halifax Dockyard.

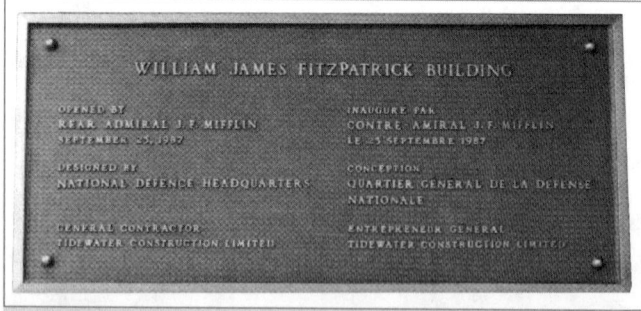

Memorial plaque.

FR7 Bill Fitzpatrick.

CHAPTER III

ARMY FIRE SERVICE

The Early Times

The Army's fire safety bible followed the principles outlined in the British Army's *Fire Manual for Hutment Camps* and the *Army Fire Manual*.[1] A 1916 edition of these manuals placed emphasis on discipline, order, the proper training of fire piquet, constant patrol inspections and the complete readiness of equipment. These were good policies that, when judiciously applied, would be certain to reduce fire losses even in today's more sophisticated times. The effectiveness of these measures probably contributed to the belief that a professional fire department was not absolutely necessary. Financial and manpower restrictions would have also played a significant role in postponement of stand-alone professional fire departments.

As has often appeared to be the case throughout the history of firefighting in all three services, success at fire prevention often seems to have delayed new programs that would have upgraded the resources dedicated to the suppression of fire and the associated rescue function. It often took a major fire or other disaster to have the firefighting and other fire protection needs recognized.

WORLD WAR I

Zealous Enforcement

Control of fire losses on military installations during this period was in large part due to the application of discipline and the zealous enforcement of comprehensive fire prevention regulations, clearly

---

[1] Canada, *Army Fire Manual for Hutment Camps* (Ottawa, 1916), p. 7.

spelled out in the *Army Fire Regulations Manual*, 1916 edition. While these fire prevention measures might appear primitive compared to the conduct of modern DND fire departments, they proved sensible and effective in their time.

The fire crews of the era were drawn, in a large part, from GD personnel lists, but also included personnel from a variety of trades. It was part of military life in those days to expect virtually all personnel to perform additional duties outside the range of their primary responsibilities. These duties were often the most mundane of the tasks that helped the military machine function, such as helping to prepare meals and conducting security assignments.

Firefighting
The organization of manpower for firefighting may have differed slightly from unit to unit, but they would have had to conform, at least in principle, to the various regulatory manuals. *King's Regulations for the Canadian Army* and *The Fire Services Manual for the Canadian Army* afforded the cornerstones for day-to-day routines. As the size of the installation dictated the type of firefighting equipment provided, the operation and deployment of fire equipment and vehicles differed from unit to unit. In many instances, the distance from the nearest community with an organized fire service was also a factor.

BETWEEN WORLD WARS I AND II

Budgetary Restrictions Dominate
The Army Fire Service experienced an extended period, as did the Navy and the Air Force, when stringent budgetary restrictions drove all other considerations. Army authorities, however, were more conscious of the dangers involved in overlooking fire safety. They sought to compensate for the lack of a full-time professional fire department by continuing the tradition that featured the energetic enforcement of fire-safety regulations. Every camp had a fire committee that organized a fire prevention program, framed fire orders, and delegated inspections of fire equipment and the conduct of fire drills.[2] The Regulations for the Army Fire Services defined the responsibilities of the fire piquet and fire parties in clearly stated detail.

Firefighting Resources
Firefighting operations were keyed to the use of hose carts, fire extinguishers, and buckets as the most effective means of getting water on a fire. During the period just before the Second World War, the fire services of the three military services were in a rudimentary state. All three were lacking both personnel and material resources, as well as training programs crucial to the development of the available personnel. Budgetary restraints and lack of manpower provided substantial roadblocks to the advancement of the Fire Service. Furthermore, the inability of the leadership to appreciate the long-term effects of fire on military operations complicated this situation. Perhaps more importantly, the fire services did not have adequate representation to advance their cause through the corridors of power in each of their respective headquarters.

The impact that major fire losses could have on military plans and operations did not become apparent until the start of the Second World War. Meanwhile, senior civilian fire officials were left with the difficult job of convincing military leaders of the potential for losses of vital war material and the possible negative effects on the war effort. A constant flow of depositions

---

[2] Canada, Army Regulations for Army Fire Services, 1934.

from professionals such as the Dominion Fire Commissioner and the Ontario Fire Marshal would eventually convince them.[3]

WORLD WAR II

Fire Protection—Starting from Scratch

In the short term, until trained firefighters became available, the Army placed reliance on fire parties (fire piquet). This system had little firefighting capability, but was very effective in providing procedures and orders for the rapid evacuation of buildings and preparing well-organized personnel muster stations. Apart from the initial first-aid firefighting efforts, the primary fire suppression activities typically fell to the local fire department. Thus, the Army Fire Service at the outset of war was practically non-existent as a stand-alone organization.

Initial Headquarters Organization

The outbreak of war brought many changes to the Works[4] component of the RCE. Lieutenant Colonel E.J.W. Akins became the Director of Works and Construction under Engineering Services. Within this directorate, there were separate sections for works and buildings, fortifications, design, lands and fire protection. As the lands function became more heavily involved in the acquisition of properties and accommodation, it was linked with the fire protection engineering function.

Fire protection activities were rapidly expanded during 1941, resulting in the appointment of Major E.J. Desjardin of the Specially Employed List as Assistant Fire Marshal. In 1942, Lieutenant Colonel O.L. Lister, also from the Specially Employed List, became Land Forces Fire Marshal.

Developing a Firefighting Force

The outbreak of the Second World War triggered a rapid expansion of the Army Fire Service, an enormous undertaking that included the recruiting, training and placement of large numbers of personnel, as well as large purchases of mobile and static fire equipment.

Army Fire Service doctrine involved recruiting personnel with municipal fire department experience and appointing them camp fire chief or deputy fire chief. Others, without experience in firefighting, were given a brief course in 1941 and 1942 at either of the Army's firefighting training schools that were situated at Camp Borden, Ontario, and Camp Chilliwack, British Columbia. Graduates were then assigned to a fire department in one of the military installations where they underwent additional on-job training. By war's end, the Army Fire Service had peaked at 1,500 personnel, distributed throughout the various military districts in Canada. This was a force of considerable consequence, especially when viewed from its modest beginnings.

Manpower Establishments

The manning of fire brigades serving the Army camps was based on the number of soldiers living in the camp. The established scale for the strength of a fire brigade followed simple guidelines, such as 800 to 1,000 personnel requiring one sergeant and one private firefighter (supplemented by fire piquet). A camp population of three to four thousand demanded one sergeant, one corporal and five privates, until an optimum of one sergeant, two corporals and ten privates was reached. Included in the equation was a plan that allowed one additional firefighter for every increase in population of a thousand personnel. Finally, a 25% overage factor was permitted for leave, courses and illness.[5]

---

[3] Captain George Cowan (1952–80), *History of the Canadian Military Fire Service* (unpublished, 1990).
[4] The Works component of Engineering Services was responsible for military infrastructure.
[5] *Canada, Army Regulations for Army Fire Services*, 1934, p. 9.

Reorganization at Army Headquarters

Into 1942, the Army Fire Marshal's Office, from an administrative perspective, was attached to the RCE while the staff remained part of the organizational strength of the Royal Canadian Army Service Corps (RCASC). Changes to this awkward arrangement were proposed in late 1942[6] to allow the Fire Marshal's Office to become an integral part of the RCE organization. The recommendation met with approval, and for a brief time, the Army Fire Service became part of the RCE. This organizational structure was short-lived, since shortly thereafter a change took place involving the Quartermaster Branch, and the Fire Marshal's Office again became part of the RCASC.

Although the Fire Marshal's Office was part of the RCASC, fire inspectors were assigned from Headquarters to work in the District Engineering Office in the military districts. The confusion that resulted from the frequent changes was alleviated to a certain degree by the perseverance of the fire inspection staff. These dedicated few who soldiered on regardless, in what must have been an administrative nightmare, must be admired.

Unique Characteristics

In many ways, the Army Fire Service was unique. For instance, they had the foresight to build fire vehicles to their own specifications and to develop innovations peculiar to their own distinct requirements. They also were able to maintain their own special identity within the land forces system, despite some agonizing organizational upheavals.

During the final meeting of the WSFPC, on March 5, 1942, prior to being replaced by the JSFC, Lieutenant Colonel Lister, the army fire marshal, announced that his service was developing a prototype fire truck in the Army workshops that would become the standard for all future Army fire pumper trucks. This initiative resulted in the production of the G 000-666 standard fire pumper that remained in use with the Army Fire Service until well into the post-war era. At this same meeting, Lieutenant Colonel Lister gave notice that he was about to appoint soldiers with previous fire department experience to the camp fire brigades. He concluded by announcing that he fully expected the Army Fire Service to increase to between 500 and 600 personnel over the next few months.

The Fire Marshal's Office

Limited in size, like many wartime organizations, the staff of Army Fire Marshals Office was expected to shoulder responsibilities that were extensive and demanding, such as:

- providing fire-safety policy and fire regulations for the Army;

Army Fire Marshal's/Fire Chief's Workshop—1984
Back Row: FR5 Cam Burelle, FR4 Ed Luce, FR5 John Beers, FR5 Eugene Gondek, FR5 Bob Bates, FR3 Rocky St-Jean and D/CFM Stony Bourque.
Front Row: FR5 Ron Rockey, FR5 Keith Scanes, CFM Gerry Berube, FR5 John Bellefleur and FR3 Bill McCroary.

---

6 Memorandum (HQ 48-1-39 December 10, 1942).

- developing standards for the provision of fire-protection equipment;
- providing advice on fire loss-limiting engineering, spatial separations between buildings, fire-protective materials, building design and operations therein;
- developing specifications for fire alarms, fire-suppression systems and fire apparatus;
- conducting boards of inquiries into fire losses; and
- preparing fire loss statistics and periodic reports on fire-service matters.

NORTH ATLANTIC TREATY ORGANIZATION

Addressing the Commitment

A period of growth, which began about 1950, was keyed to Canada's participation in the North Atlantic Treaty Organization (NATO) that, among other things, resulted in the opening of Army camps in the northern part of West Germany in the early 1950s and to the development of a fire service staffed largely by German nationals. The deployment included an Army Brigade Group that was to form an element of the British Army of the Rhine. A works company was created in 1951 to reconnoitre the forts that had been set aside for Canadian occupation. They were to form a component of the advance party for 27 Canadian Brigade Group (CBG). Eventually, they would oversee the renovation and construction of accommodations located close to the villages of Soest, Werl and Iserlohn.

After considerable expenditure of time, effort and money by the Army Engineers, the construction was completed in 1953. The hiring of a totally civilian engineering support section from the local populace was thought unwise; consequently, an engineer works service was incorporated into the CBG. Part of its responsibilities included fire protection services.

Army NATO Units

The first rotation in 1951 consisted of units of 27 Canadian Infantry Brigade (CIB). Based on units from No. 1, Canadian Infantry Brigade Group (CIBG) was formed in 1953 to replace 27 CIB which, in turn, was replaced by No. 2 CIBG in 1955, followed by No. 4 CIBG in 1957. The units were quartered in a series of forts as outlined below until they moved to Lahr, West Germany, in 1970.

Fort Anne
Lord Strathcona's Horse, Werl, West Germany (Armour and Field Ambulance).

Fort Beausejour
Royal Canadian Dragoons, Iserlohn, West Germany (Armour).

Fort Chambly
Service Support Units, Soest, West Germany.

Fort Henry
Headquarters, Canadian Mechanized Brigade Group, Stockum, West Germany.

Fort McLeod
Princess Patricia's Canadian Light Infantry, Hemer, West Germany.

Fort Qu'Appelle
SSM Battery. This fort in Iserlohn, West Germany, was not occupied until 1968. (Surface-to-Surface Missiles)

Fort St. Louis
Royal 22^e Regiment, Werl, West Germany (Infantry).

Fort Victoria
Royal Canadian Engineers, Werl, West Germany.

Fort Prince of Wales
Royal Canadian Horse Artillery, Deilinghofen, West Germany.

Fort York
Royal Canadian Regiment, Stockum, West Germany (Infantry).

A Novel Approach—Fort Prince of Wales[7]
In the fall of 1957, Subaltern Mike Calnan, of the Royal Canadian Horse Artillery (RCHA), was serving with the regiment in Winnipeg prior to proceeding to Germany to, among other things, serve as Unit Fire Prevention Officer for Fort Prince of Wales. Prior to his departure he was sent on one of the first fire-prevention courses at the Royal Canadian School of Engineering at Camp Chilliwack, British Columbia. During his first year in Germany he inspected, drilled and helped train the German civilian fire detachment in Fort Prince of Wales. The camp consisted of some 50 single-storey buildings, the largest being the arena and the churches. The fire detachment, supported by the regimental guard, had little to challenge their capabilities, but turned out once a month for camp fire drills and the testing and inspection of all fire equipment. Years later, Colonel Calnan returned to Fort Prince of Wales as the proud new Commanding Officer of 1 RCHA, only to find that it was required to downsize and move to Lahr.

The Fire Service in West Germany
The fire service that evolved in West Germany comprised of 51 German nationals, an RCE fire inspector as fire chief, and a bilingual German national as deputy fire chief. The ability of the deputy to speak both German and English was an important asset. Each army fort had a fire department housed in a small fire hall. The basic fire hall design consisted of an office, a single-bay garage, a small maintenance shop, a kitchen/lounge and sleeping accommodations for four firefighters.

Some 16 years later, a government decision to co-locate the Army units with the Air Force units in southern West Germany, mainly in Lahr and to a lesser degree Baden, where a fire service was in place at both locations, resulted in the disbanding of the Army Fire Service in West Germany.

CWO Les Duguid with German civilian firefighters in West Germany.

Apparatus
Volkswagen manufactured the first generation of fire trucks that were purchased. These vehicles came equipped with a small pump capable of delivering approximately 2275 L/min / 500 G/min, a quantity of 65 mm / 2½ inch hose and 38 mm / 1½ inch hose, as well as an assortment of firefighting tools. A modest piece of fire apparatus that, nevertheless, was quite effective in its forecast role.

---

[7] Memoirs of Colonel Mike Calnan (1950–87), Royal Canadian Horse Artillery, February 2004.

Mutual Aid
Mutual aid, whereby the fire service of the CBG interacted with the fire services in the local communities to support one another in time of need, quite similar to the mutual-aid agreements commonly engaged in Canada, was an important component of the fire-protection services. This commitment to participate in the protection of surrounding area in this manner proved to be an effective method of cementing relationships with the local populace.

Murphy's Law
Murphy's Law, the facetious proposition that if something can go wrong, it will, is often proven to be as irrefutable as the laws of physics. How it came into force is based largely on ironic humour. In 1949, Edward Aloysius Murphy (1917–91) was a United States Air Force (USAF) aeronautical engineer working on rocket-sled trials at Edwards Air Force Base in California. The tests were designed to see how much acceleration a jet pilot could stand. After one trip down the rail tracks at 600 mph / 960 kph, a gauge that was supposed to measure the strain on the rider read zero; a technician had wired it backward. An exasperated Murphy said, "If there was any way to do it wrong, that guy would do it."

Colonel John Stapp, USAF, the guinea pig who was risking his life in the tests, modified Murphy's statement to read, "If it can happen, it will happen." This saying remained an inside joke among the test team until Colonel Stapp used it at a news conference on January 5, 1950, to discuss the now successful program. He said, "We are great believers in Murphy's Law." From that point on, it has been in the public domain. Over time the description, if there are two or more ways to do something and one of those ways can result in a catastrophe, then someone will do it, also became a common interpretation of the law.

Colonel Stapp, applying some gallows humour, developed the following rule as an extension to the law: "The universal aptitude for ineptitude makes human accomplishment an incredible miracle. Mercifully, the name of the bumbler, upon whose actions the law was founded, has not been recorded."

Murphy's Law—Case One
In 1962 or 1963, Flight Lieutenant John Cowell, Fire Marshal for No. 1 Air Division, with headquarters in Metz, France, accepted an invitation to visit the Army facilities in the Soest area. The firefighters at the Army camps, with foreknowledge of his impending arrival, were primed to perform at their best and possibly amaze the visiting dignitary with their firefighting proficiency. The fire marshal duly arrived at Fort Henry and asked for the opportunity to observe a hose-and-ladder drill. After selecting a suitable building, he then outlined the type of drill he envisioned. In due course the drill got underway, consisting of laying two hose lines, charged with water, and then setting a ladder against a nearby building to gain access to the roof. A fairly straightforward evolution and, to the crew's credit, everything was going quite well until it was time to raise the ladder. At that point the seemingly confident crew unloaded the extension-type ladder from the pumper and placed it upright near the building, intending to extend it to the roof. But try as they might, the ladder refused to extend despite their persistent pulling on the lanyard. Realizing the cause of their dilemma the fire marshal walked over and, much to the chagrin of the fire chief, explained that the ladder had been placed upside down. Not exactly a banner day for the fire department, but there again, these things happen from time to time, especially during drills. In the best traditions of Murphy's Law, if something can go wrong, it will do so when you are trying to impress someone.

Murphy's Law—Case Two
Following the exercise at Fort Henry, the fire marshal and the hosting party travelled to Fort McLeod. On arrival there, he decided to have a close look at one of the German-built pumpers, which he had not seen previously. The accompanying group followed until he stopped and opened one of the vehicle's compartment doors. It was the compartment that housed all the important pump controls and levers. On opening the door, he found a lawn mower jammed into the compartment, effectively denying anyone access to the pump controls. There was no way to use the pump without removing the mower. This might appear to be a simple task that could be done quickly; such, however, was not to be the case. The mower was tightly jammed in place and resisted their best efforts and the influence of the ample application of colourful language for some time before extraction was achieved.

It turned out that the mower was going to be used to cut the grass around the fire hall. When the crew picked it up from the Construction Engineering (CE) section they discovered there was no place to carry it, except in the compartment. On returning to the fire hall they found coffee break in progress which they all settled down to enjoy, thus leaving the mower inside the compartment. Quite likely the vibration of the vehicle over the roads from the CE section to the fire hall caused it to become wedged in place. Proving once again that if something can go wrong, it eventually will, and will do so at the most inopportune time.

The Real Picture
The persistence of Murphy's Law aside, the guest fire marshal found that the fire departments at the forts were well run and in a good, operationally-ready state.

CHAPTER IV

AIR FORCE FIRE SERVICE

WORLD WAR I

The Beginnings of a Fire Service

The exact date when the formation of an air force fire service took place is somewhat obscure, although the first phase of the history of the military fire service has a direct association with the development of British Flying Services in Canada at Camp Borden in 1917. The Camp's isolated location and the need to be self-reliant with regard to fire protection may have been instrumental in stimulating the beginnings of a Canadian military fire service that comprised two distinct and separate functions: aircraft-crash responses and structural firefighting.

A photograph (circa 1918) of uniformed military firefighters illustrating personnel on duty at Camp

The first Air Force fire hall (1918), Fire House, Camp Borden, with a Model "T" Ford crash tender. The Fire House was used by the RFC and the RAF before being handed over to the CAF.

Borden confirms the existence of an organized military fire brigade. Six years later, on April 1, 1924, the RCAF came into being when the prefix "Royal" was added to the Air Force title, and King's Rules and Regulations were promulgated.

BETWEEN WORLD WARS I AND II

Carrying On
Reductions of personnel were the order of the day between the two world wars. Nevertheless, when sufficient resources were available during this period, the Air Force conducted many useful services, including photographic surveys of the north, forest-fire patrols and even delivering treaty money to the Indian reserves. As the reductions continued, coincident with the imposition of further financial restraints, all these services were cut. The Air Force barely survived. Then, in 1935, the military appropriation was boosted and the manpower establishment was allowed to increase substantially.[1]

Dramatic Development—A Separate Entity
During this period, the Air Force was administered by a senior air officer who was responsible to the Chief of the General Staff of the Army. This situation existed until 1938 when the Air Force became a separate organization with a Chief of the Air Staff (CAS) directly responsible to the Minister of National Defence. Air Vice Marshal Croil became the first CAS on November 15, 1938.[2]

Fire Protection—The Early Years
From 1920 until the start of the Second World War, Air Force stations were in large part responsible for developing their own fire protection. Airmen were assigned fire piquet duties on a regular basis to help meet those responsibilities. Only three pre-war Air Force stations had fire trucks; the remainder relied on hose-reel carts. These were augmented by fire extinguishers and the occasional 40 G / 190 L wheeled-unit foam extinguisher that were customarily reserved for locations where quantities of flammable liquids were stored.[3] Local municipal fire services were called upon when needed to supplement a station's firefighting capability.

The fledgling Air Force operated stations at Vancouver, British Columbia; High River, Alberta; Winnipeg, Manitoba; Camp Borden, Ontario; Ottawa, Ontario; and Dartmouth, Nova Scotia.[4] Although there had been a dedicated airfield and structural fire service during the First World War, from 1924 until 1939, there was no organization within the RCAF that could be called a cohesive and structured fire service.

Appointing a Fire Chief—A Unique Practice
It was the practice of the commanding officer of an Air Force station to appoint a station fire chief from his staff of commissioned officers. One appointee of note was Flight Lieutenant R. Slemon, who later attained the rank of air marshal. Describing fire drills at Camp Borden in the 1930s, Charles Armstrong wrote:

> I was a member of the fire crew for a number of months. My duty was to connect the hose to the fire hydrant and turn on the water. When I was on the fire crew our fire chief was Flight Lieutenant Slemon, later to become an Air Marshal and went on to become Chief of the Air Staff.

This appointment was regarded by most officers of the day as an extra duty and, therefore, its award generated

---
1   Christopher Shores, *History of the Canadian Air Force* (Toronto: Bison Books, 1984), p. 18.
2   W.A.B. Douglas, *The Creation of a National Air Force* (University of Toronto Press, 1986), p. 51.
3   *Canada, Army, Army Fire Manual Issued with Army Orders* (Ottawa, 1917), Appendix D.
4   Madeleine Stace, *History of the Construction Engineering Branch*, p. 5.

little elation from the recipient. Although the assignment of commissioned officers to head the station fire department with negligible firefighting experience could be taken as an indication of the low importance that was assigned to the military fire services of the time, it is more probable that the practice was based on the philosophy that a commissioned officer had to bear final responsibility in any given field. In time, the importance of fire protection issues gained greater recognition in higher military circles, which, in turn, led to the development of a professional fire service.

Continuing with another portion of Charles Armstrong's letter:

> Following the large cut in strength in 1932 the Station was closed for the month of July. One fire crew was on duty for 24 hours a day for two weeks and then another crew took over for the next two. I was on the first two weeks; our mess had closed down so we had to eat at the Army Signals Mess. We had to stay with or near our truck so the normal routine was to rise in the morning, get cleaned up and dressed, then all on board the fire truck for the trip to the Signals Mess. This procedure of taking the truck to the mess was repeated for each meal. During the afternoon we could take the truck down to the swimming pool for a dip, or to the ball diamond for a ball game. We had a nine-man crew and thus we could make up one full team, no spares, no coaches, but it did help pass the time and made good entertainment for us.

This gives an insight into the state of the Fire Service during the post World War I period.

Large Fires—Little Impact

There were several fires of singular importance during this period. In February 1930, the officers' and airmen's messes at Camp Borden burned to the ground. These fires were closely followed by the destruction of a barrack block. Fortunately, no lives were lost. The destruction wrought by these fires was substantial, yet did not prompt immediate improvements in the quality of fire protection.

Fundamental Changes

The first fundamental change within the Air Force Fire Service took place on August 26, 1937,[5] when the chief aeronautical engineer, Group Captain E. W. Stedman, advised the Senior Air Officer that his department had found it impossible to keep abreast of developments in firefighting technology and training. He made two recommendations:

- that the Air Force employ a firefighting expert on a full-time basis to direct the total training of Air Force Fire Service personnel and to act in an advisory capacity in the design and purchase of firefighting apparatus; or failing that,
- employ a fire expert in a consultant capacity.

In response to the second recommendation, Lieutenant Colonel L.R. LaFeche, Deputy Minister of National Defence, held a meeting with Mr. Finlayson, Superintendent of Insurances, and Mr. Grove Smith, the Dominion Fire Commissioner, to discuss the fire-service situation within the Air Force. This meeting proved to be a landmark as far as the Air Force Fire Service was concerned, as from it would spring the changes necessary to accommodate wartime expansion.

---

[5] Directorate of History, HQ File 902-1-31 (Ottawa, 1942).

WORLD WAR II

Setting Up a Fire Service Organization

As a result of the discussions between the Dominion Fire Commissioner and the Superintendent of Insurance, the Commissioner's Office agreed to appoint a suitable fire-protection specialist to inspect Air Force installations and to study the adequacy of existing fire-protection services. As a consequence of these inspections, it was recommended that the Air Force appoint an individual to take over the administration of the Air Force Fire Service. Air Force Headquarters then sought to recruit a person to oversee the management of its fire-protection services. A circular was issued to advertise for an acceptable candidate through the Civil Service Commission. The circular read as follows: "Fire Prevention Engineer, male, Department of National Defence, Ottawa $2,220 per annum."[6] This was a vital step in the development of the Air Force Fire Service in that it established a recognizable body within the Air Force that was accountable for its continued growth and operational efficiency.

In 1939, Mr. L.J. Bishop, a member of the Ontario Fire Marshal's Office, volunteered for the advertised position. His credentials were favourably received, and he was accepted into the Air Force. Also accepted into the Air Force was a Mr. P.S. Snarr who had previously been employed by the Pyrene Manufacturing Company as a fire-protection engineer.

Both men were given their assignments, and in 1940, Flying Officer Snarr was assigned to AFHQ as the Fire Prevention Officer, while Flying Officer Bishop was appointed Command Fire Prevention Officer at No.1 Training Command Headquarters in Toronto. He remained there until January 1943, when he was posted to Eastern Air Command.

Flying Officer Snarr assumed an important role during these formative years. He attended the initial organizational meeting that resulted in the formation of the WSFPC on April 29, 1941. He later became the Air Force representative on this committee until November 1941 when, because of deteriorating health, he was replaced by Flying Officer (later Squadron Leader) J.E. Ritchie, who had recently left the Ontario Fire Marshal's Office to take up the position.

The Ontario Fire Marshal's Office played a major role by supplying experienced fire officers to spearhead the building of the Armed Forces Fire Services. Their willingness to serve the forces by generously sharing their manpower resources was truly commendable. Each skilled individual who left civilian employment to join the military took on a significant increase in his personal workload.

The Japanese Threat

As a result of the United States' concern for the Japanese threat, in 1940, the Canada–US Permanent Joint Board of Defence recommended that an air route be built from Edmonton, Alberta, to Fairbanks, Alaska. The route was required in order to allow air traffic between Edmonton, and Fairbanks, a distance of approximately 2,000 miles / 3200 kilometres, as the aircraft of the day did not have sufficient range to make the journey non-stop and also to permit unscheduled landings in the event of in-flight problems.

The proposed route was one that had already been tentatively decided upon by the Canadian government to counter the Japanese invasion threat. Named the Northwest Staging Route (NWSR), it comprised a series of airfields with intermediate landing fields that constituted an aerial highway that would provide a

---

6   Directorate of History, HQ File 921-1-31 (Ottawa, September 5, 1939).

means of rapid deployment for troops and supplies. In due course, firefighters were called upon to man these northern outposts. NWSR in effect, became an aerial version of the Alaska Highway.

Work on the initial sites began in February 1941, and No. 4 Training Command assumed responsibility for the operation and maintenance of the NWSR in July 1942. By October of that year, firefighters were already moving into their new NWSR fire halls, complete with standard-issue pumper and crash truck, to support this operation. The need for firefighters increased with the increase in air operations, and a steady flow of replacements headed north. The additional firefighters were well received by personnel already on site.[7] Living and operating conditions were extremely harsh on these northern airfields, with firefighters hard-pressed to keep their vehicles operating in the extremely cold winter months. The more remote the installation, the more self-reliant the fire department had to be. Training station personnel and the scaling[8] of the various buildings took up a considerable amount of their time.

Organizational Change
For some time, the Fire Service was aligned with the Directorate of Supply Administration, with an establishment of one flight lieutenant, one warrant officer II, and three clerks. When the Air Force restructured its supply organization in 1942, the fire protection responsibility moved from the Supply Administration organization to become part of the Works and Buildings (CE) Division. The Works Fire Protection Officer (Fire Marshal) reported to the Director of Works and Buildings. The Fire Service has basically remained within this organizational framework.

Endorsement of a Leader
An interesting footnote relating to the Air Force Fire School comes from a 1942 article printed in the RCAF Station Trenton's *Trenton Contact*. The article in part states that, "There are four instructors representing sixty-seven years of collective firefighting experience, headed by Flight Sergeant T.H. Matthews, a former member of the Toronto Fire Department, his assistant Sergeant C.W. Pollard a westerner from the Regina Fire Department, third in charge Corporal C.A. Thebarge, a past Lieutenant in the Ottawa Fire Department, and finally Corporal F. Wilson. All four claimed to miss the presence of their former Chief, Flying Officer W. McCallum, now stationed at Winnipeg and for whom they have great admiration."[9] This was a splendid endorsement from a group of professionals who, in their own right, had considerable background and experience professional firefighters.

POST WORLD WAR II

Exodus of Personnel
The end of the war witnessed the departure of large numbers of firefighters from the ranks of the armed forces. This unparalleled exodus created a shortage of firefighters needed to carry out fire inspections and to provide effective support for flight operations and firefighting and rescue operations in general. After suffering a series of costly fires, the Air Force initiated a recruiting program aimed at increasing its number of firefighters. As a consequence, the fire school at RCAF Station Mountain View reopened in late 1946. This is chronicled in more detail in Chapter VI. Although the need for expanding the Fire Service was underway prior to the demands inherent in Canada's commitments to NATO and North American Aerospace Defence (NORAD) coming into effect, much remained to be done.

---

7 Major Phil Brown (1947–75), *The Royal Canadian Air Force Fire Service 1939–75, a Subjective History* (unpublished).
8 See Appendix B for a definition.
9 *Trenton Contact*, March 1942, p. 7.

The Struggle for Leadership

The immediate post-war period witnessed several changes of personnel, particularly in the hierarchy of the Air Force Fire Service. The release of Squadron Leader Richie left the senior staff position at AFHQ vacant. His replacement was Wing Commander W.D. Martin, a civil engineer with the Directorate of Works and Buildings. During this period, Warrant Officer First Class R. Armour was commissioned as a fire officer and posted to AFHQ as Senior Works Fire Prevention Officer. In spite of receiving his commission, he was persuaded by a fire vehicle manufacturer to forego his military career and join them in a civilian capacity. This sudden departure resulted in Warrant Officer B. Quinn being selected as his replacement and posted to AFHQ. He was subsequently commissioned with the rank of flying officer. He served for many years before retiring with the rank of wing commander.

At the Fire Departments

Every Air Force flying station had at least one structural pumper and one crash tender. Normal deployment during flying operations had the pumper on standby in the fire hall and the crash truck positioned at the base of the control tower. The larger airfields that had a satellite landing field were provided an additional crash tender to be dispatched to the satellite site when flying was in progress. Hangar-line personnel often provided the crew for the extra vehicle, chiefly because of their knowledge of the particular aircraft. Their assistance often proved invaluable. The catch-phrase for the firefighters who spent their days in front of the control tower was monotony, with the only activities watching routine flying operations, or responding to an occasional incident. It was truly a trying task.

Manning levels of each department varied, depending on the responsibilities and operational activities of a particular installation. For example, airfields dedicated to flying training had a flight sergeant appointed as the fire chief, with a sergeant as deputy fire chief and between 16 and 18 crewmen. The manpower was divided into two shifts, each working 24 hours on duty / 24 hours off duty. Where there was room, all firefighters were expected to live in the fire hall. The rationale was that, if a firefighter was in the fire hall when an alarm sounded, regardless of his duty status, he was automatically presumed to be on duty. During emergencies extra crew members were welcome additions.

Most wartime fire halls of the Air Force were a standard design with only minor modifications to satisfy local requirements. Although the buildings were quite austere, it was a practical solution that saved time and money as the construction cost of one of these halls was $10,500.[10] Wartime fire halls embodied a two-bay vehicle garage, an office where the coded alarm panel with a tape punch register was located, a combined workshop and hose drying room, a firefighter dormitory, a common shower and washroom, a recreation lounge, and a combined heating and laundry room.

The daily routine for the airfield crash crew consisted of standing-by for flying operations by pre-positioning their crash truck at the foot of the control tower or other position of advantage. When assigned to duty in the fire hall, the crew typically worked on maintaining fire extinguishers, repairing equipment, painting, conducting fire inspections and training evolutions. They would normally be expected to handle most structural fires independent of the crash crew, if flying operations were still in progress. This type of activity has been the lot of the firefighter since the very first organized fire brigade stood watch. It was not until the early years of the Cold War that fire safety education,

---

[10] Madeleine Stace, *History Of the Construction Engineering Branch*, p. 18.

fire loss-limiting engineering, and enforcement of fire regulations took their place along with operations to form the "Four Key Factors of Fire Protection."

The COLD WAR

International Tension

The intense and acrimonious rivalry between the Western nations and the Soviet Union resulted in increases in manning levels, which created a shortfall of trained personnel to fill vacant positions. Canada, like its allies, began to expand its military forces at home and in Europe as a consequence of the deepening of the Cold War. Senior staff at AFHQ attempted to overcome the shortage by recruiting newly released wartime firefighters. Before the backlog was fully addressed, trained firefighters from as far afield as the British Isles were recruited, some of whom rose to prominence within the Air Force Fire Service. In the early 1950s experienced firefighters were also recruited from civilian fire departments in Canada. Many came from the City of Ottawa and were automatically promoted to sergeant rank, to the considerable consternation of serving corporals who were qualified and eligible for promotion.

By 1954, Canada had an air division of 12 fighter squadrons and an army brigade based in Europe. Further motivation for expansion of the Armed Forces had occurred in June 1950 when the Communists attacked South Korea. The reaction of the Allies in expanding their forces included upgrading firefighting capabilities, and thus the Fire Service continued its overall development.

Materiel

The rapid expansion of the Armed Forces, coupled with the neglect of the inter-war period, had left the Fire Service virtually bankrupt with respect to portable and mobile fire equipment. An impressive procurement program was initiated to remedy the situation. The Air Force alone purchased over 100 G10 structural pumpers and 250 crash trucks. In addition, several hundred hose-reel carts and thousands of fire extinguishers were also procured. Two sizes of hose were acquired: the 2½ inch /65 mm and 1½ inch / 38 mm rubber-lined, double jacket, cotton-covered hose. These were excellent hose, that would serve the Fire Service satisfactorily for many years.

The 1½-inch / 38 mm hose was delivered with the same number of threads per inch/cm as standard pipe threaded couplings. There was one key difference: pipe couplings were tapered to affect a watertight seal between the male and female, simply by screwing them tightly together, which required considerable torque. This design was too slow to use for firefighting operations; consequently, the hose couplings were straight cut to permit easy connections and used compression of a rubber gasket between the male and female coupling to form a water-tight seal. Unfortunately, the 2½ inch / 65 mm hose was delivered with some 27 different thread specifications, in keeping with the mix that existed in the various geographical locations.[11] This subject is addressed in more detail in Chapters VIII and IX.

Leadership Foundation

In 1952, ex-Flying Officer W. (Bill) McCallum, then a senior officer with the Toronto Fire Department, was approached with an offer of a commission. The offer was made in order for him to take over the position of Air Force Fire Marshal (AFFM) with the rank of squadron leader. McCallum, who was already planning to retire from the Toronto Fire Department at age 60, happily accepted the offer. In 1956, he was promoted to the rank of wing commander. He was an amazing individual with an exceptionally interesting career

---

11 Memoirs of Major Walter Sinclair, *The RCAF Fire Service* (unpublished).

history. His many achievements include: World War I veteran, youngest warrant officer in the Canadian Overseas Army, senior officer in the Toronto Fire Department, World War II fire prevention officer, and finally, fire marshal for the post-war Air Force.
In 1988, Wing Commander Bill McCallum passed away in Toronto, at age 93.

Bilateral and Multilateral Agreements
In April 1949, Canada signed the North Atlantic Treaty which committed member nations to come to the defence of any member who came under attack. The Armed Forces' rebuilding program began in earnest.

In 1954, Canada and the United States entered into a bilateral agreement to consolidate their air forces for the defence of North America under an organization called the Canada–USA Regional Planning Group. Initially, this was a NATO Command before being formalized in 1958 under the title of NORAD.

NATO

Composition of the Air Force in Europe
In October 1951, the first unit (410 Squadron) of the Air Force left Canada aboard the aircraft carrier HMCS *Magnificent* to begin fulfilling the Canadian air commitment to NATO. Within two years, a headquarters formation, four fighter wings and an air materiel base had been established in Europe.

No. 1 Air Division Headquarters, located at Metz, France, was the administration and operations command centre for the Air Force contingent in Europe.

The RCAF No. 30 Air Materiel Base at Langar also served as the European terminus for transatlantic flights by the North Star aircraft and was home to the venerable Bristol Freighter aircraft[12] of 137 Flight, which provided logistical support freight service to the Canadian units on continental Europe. The arrangements for operating the airport in Langar, although quite similar to what was in place in a number of airports in Canada, were unique relative to other airports operated by the Air Force in Europe. Although the Air Force provided the key essential services, such as air traffic control, navigational aids, firefighting and rescue and airfield maintenance, it was a shared airport. Most of the Air Force installations were on one side of the airport, with a civilian industrial complex on the opposite side.

No. 1 Fighter Wing was initially composed of Squadrons 410, 439 and 441. This fighter wing was first located at North Luffenham, Rutland, England, but moved to Marville, France, in the spring of 1955. No. 2 Fighter Wing was located in Grostenquin, France, and originally included Squadrons 416, 421 and 430. No. 3 Fighter Wing was located at Zweibrucken, Germany, and originally was formed of Squadrons 413, 427 and 434. No. 4 Fighter Wing was located at Baden-Soellingen, Germany, and included Squadrons 414, 422 and 444. In the late 1960s the Air Force relocated from Marville to Lahr, West Germany. At this point, Lahr became the terminus for both passenger aircraft and cargo aircraft plying the transatlantic routes. As well, it was the hub-airport for air transport flights in support of Canada's various commitments on behalf of the United Nations, most particularly Cyprus and the Middle East.

With an air division deployed in France and West Germany, and an army brigade group located in the northern part of West Germany, Canada provided a significant military presence within NATO in Europe.

---

[12] Christopher Shores, *History of the Royal Canadian Air Force*, p. 16.

Profile of the Fire Service

As a component of 1 Air Division Headquarters (ADHQ) at Metz, the Fire Marshal's Office was administered by a flight lieutenant assisted by a warrant officer. The Headquarters moved to Lahr in the late 1960s and, shortly thereafter, became the headquarters for the Canadian Forces in Europe.

These European installations needed first-rate fire protection and generated a new demand for trained firefighters. Many ex-RAF firefighters were recruited through the office of the Canadian Joint Staff (CJS) located in London, England. They were inducted into the RCAF as trained firefighters, or awarded provisional trade qualification groupings.

Initially the fire chief position was established as a warrant officer second class with a flight sergeant as deputy chief, and with sergeants as crew chiefs at 30 Air Materiel Base, Langar. Each wing had a flight sergeant fire chief while the deputy chief and crew chiefs were sergeants. At the Metz site the fire chief was a sergeant and the crew chiefs were corporals. As events unfolded and the risk increased, fire department establishments increased markedly. So too did the rank structure. All fire chiefs were commissioned FSOs, with master warrant officers as deputy fire chiefs and warrant officers as crew/platoon chiefs. During the peak years when nuclear munitions were involved there was, at any given point in time, in excess of 200 military firefighters in Europe.

Equipment

Much of the firefighting equipment was sent from Canada to Europe, but there was still extensive training required for firefighters arriving from Canada. The training was necessary because considerable quantities of equipment and vehicles, particularly fire vehicles, had been purchased in Europe and had unusual operating controls. Initially the fire vehicle fleet was composed of a Bickle Seagrave G11 pumper, a G21 Thornycroft major foam vehicle, and an International G13 dry chemical truck.

Unique Operations Demands

The prevailing conditions in Western Europe necessitated change in the way firefighting operations were carried out. The picture differed greatly from the armed forces posture in Canada. The need to be as fully prepared as possible to thwart an attack by a hostile force was of paramount importance. It was necessary, for example, to carry small arms and respirators in addition to the normal firefighting equipment, when exercises structured to simulate wartime conditions were in progress. It was a cumbersome

No. 1 Air Division Fire Chiefs' Conference held at No. 3 Wing, Zwibrucken, in April 1957. F/S Ivan Folkerson No. 4 Wing, Baden; F/S Ken Stokes No. 1 Wing, Marville; WO2 Ron Parker, 30 AMB Langar; WO 1 Dave Lafebvre, No. 2 Wing, Grostenquin; F/L Archie Graham, Fire Marshal, Metz; Sgt Bill Brady, Metz; and WO2 Larry Dagg, No. 3 Wing, Zwibrucken.

kit; however, it was an unavoidable fact of military life in Europe during the period when the tensions inherent in the Cold War were most threatening.

In addition to expanding established manpower ceilings, the deployment of forces to Europe gave a renewed sense of purpose to the Canadian military. The heightened sense of purpose was further intensified by the constant training operations being flown by the squadrons. These activities had an impact on all facets of station life, with the fire department being no exception.

Numerous crash-alarm responses became the norm for the firefighters based in Europe, as the Air Force policy of the time was to train as many aircrew as possible. The continual influx of new pilots meant that inexperienced pilots and high-speed aircraft often translated into frequent airfield emergency responses.

The fire department at ADHQ in Metz faced a different challenge. Immediately after exiting the fire hall on an emergency response, with a Bickle Seagrave G11 pumper, there was a very steep hill to climb. While there are no reports of its failure to make it up the hill even in icy conditions, according to some of the firefighters it really made life interesting in winter.

New Firefighting Technique[13]

In 1970, No. 1 Wing Fire Department (1 WFD), Lahr, West Germany, conducted a series of tests to determine the level of effectiveness of Class A Medium Expansion Foam (MXF) on fires in buildings. The test results were very impressive. As a consequence, 1 WFD began using foam to fight structural Class A fires in late 1970 or early 1971, long before it was generally accepted as a method of fighting structural fires in other military or civilian departments. This put the DND Fire Service in the vanguard of fire services respecting the use of foam in fighting structural fires. Application of MXF in basement fires proved extremely effective, extinguishing the fire quickly, while minimizing water damage. Occasionally, it saw application in multi-storey buildings, invariably with excellent results.

The effectiveness of the technique led to another innovation, specifically, triple combination pumpers being fitted with an integral foam tank. The addition of a 40 foam tank increased the potential usefulness of a pumper considerably, by providing the vehicle with a ready reserve of MXF for structural firefighting. It also permitted quicker application of foam to a fire situation and allowed the pumper crew the flexibility to successfully fight small flammable liquid fires, or vapour-seal the surface of a dangerous fuel spill. This innovation was adopted by many forward-thinking city and municipal fire departments.

NORAD

Purpose and General Structure

The NORAD agreement between Canada and the United States was formalized in 1958. The aim of NORAD was to create a defensive screen against the threat of an air attack by the Soviet Union. Its motto is "Deter, Detect, Defend." Clearly an early warning of any potentially hostile air traffic was the cornerstone of the system. To achieve this, a network of aircraft detection and aircraft-control radar systems was constructed in Canada by the United States and Canada. There were three lines called the Pinetree Line,[14] the Mid-Canada Line and the Distant Early Warning Line. These new units were often isolated, and all demanded fire-protection services. World events were reshaping the Canadian Forces and, with them, the Fire Service.

---

13 Memoirs of Lieutenant Colonel Lorne MacLean (1951–87).
14 See Appendix E for a listing of the stations on this line.

USAF Strategic Air Command Bases

In the early 1960s, Canada and the United States reached an agreement whereby the USAF would, in order to extend the range of their strategic bombers, operate KC97 in-flight refuelling tanker aircraft out of four northerly Canadian airports located in Frobisher Bay, Churchill, Cold Lake and Namao. The risk involved required a strong firefighting service.

USAF Strategic Air Command Bases—Staffing

The staffing of the fire departments at these airports was unique. Each was established for duty crew strength of 26 military firefighters which included three from the USAF and three from the RCAF, the remainder being Canadian civilians. The military contingent included a flying officer fire chief, a flight sergeant deputy fire chief, two sergeant platoon chiefs, two corporals and eight aircraftsmen.

The agreement required the USAF to fund the employment of civilians to supplement the RCAF and USAF firefighters. The RCAF Fire School at Borden provided a seven-week aircraft crash-rescue and firefighting course for civilians recruited for the purpose. A total of 56 recruits were trained under this plan.

A meeting of Air Force fire marshals, 1958.
Back Row: F/O Jules Debrouvers, F/O John Cowell, F/O "Hank" Webster, F/O George Palmer, F/O Wally Sinclair, F/O Phil Brown, F/O Bill Maggs, F/O Archie Graham, F/O Leo Herman and F/O Jerry Torraville.
Front Row: F/O Ron Parker, F/O Phil Barrett, F/L McFadyen, A/C Whiting, S/L Bert Quinn, G/C Ingles and F/L Walker.

USAF Strategic Air Command Bases—Apparatus

The inventory of mobile fire apparatus included a mix of RCAF and USAF vehicles. For example the establishment at Churchill included: two 011A MFVs (US); one Tanker (US); one light rescue vehicle (LRV) (US); one MFV (Cdn); and one LRV (Cdn).

USAF Strategic Air Command Bases—Termination of the Agreement

This arrangement remained in place for approximately two years. It was terminated as the result of a change in strategic policy by the USAF that ended the requirement. Most of the civilian firefighters made redundant by this policy change were able to find employment as DND firefighters, with many accepting positions as firefighters at radar stations.[15]

NUCLEAR MUNITIONS[16]

Why and How

In the late 1950s, the Government of Canada, working within the parameters of the NATO and NORAD

Air Force Command Fire Marshals' Conference, 1961
Back Row: F/L George George Palmer, F/L Wally Sinclair, F/L "Hank" Webster, F/L Phil Brown, F/L Bill Maggs, F/O Brady and F/L Jerry Torraville.
Front Row: F/L John Cowell, S/L Bert Quinn, G/C Baker, F/L Archie Graham, F/L Bill Walker and F/O Barrett.

---

15 Memoirs of FR2 Marcel Trudel, Deputy Fire Chief, CFS Gypsumville.
16 Definitions of the various terms that are used in this broad subject area are given in Appendix B.

Agreements, decided to increase Canada's defence capabilities by including nuclear warheads in its arsenal. The weapons held in Canada were solely intended for use in the air defence of North America, while the weapons at Canadian facilities in Europe were an air to serve as a deterrent against an attack on Western Europe. While the actual warheads would remain under United States administrative control until the outbreak of hostilities either occurred or were imminent, their physical safekeeping was the responsibility of the Canadian Armed Forces. The USAF and the RCAF conducted intense and ongoing programs of training respecting their safekeeping and operations procedures. Canada had three delivery systems, those being the Boeing IM-99B Bomarc surface-to-air missile, the Air-2A air-to-air type weapon for use with the CF 101 Voodoo interceptors, and the air-to-ground type used with the CF 104 Starfighter aircraft.

Nuclear-Capable Sites

The Bomarc sites were located at La Macaza, Quebec (447 Squadron), and North Bay, Ontario (446 Squadron). The stations operating CF101 Voodoo aircraft were: RCAF Station Comox, British Columbia; RCAF Station Bagotville, Quebec; RCAF Station Val d'Or, Quebec; and RCAF Station Chatham, New Brunswick. The formations operating CF104 Starfighter aircraft were No. 3 Fighter Wing, Zwibrucken, West Germany; No. 4 Fighter Wing, Baden-Sollingen, West Germany; and No. 1 Fighter Wing, Lahr, West Germany.

The time periods during which nuclear munitions were in place are as follows:

- RCAF Station Bagotville, Quebec, May 31, 1965 to April 11, 1984
- RCAF Station Chatham, New Brunswick, May 31, 1965 to March 31, 1975
- RCAF Station Comox, British Columbia, May 31, 1965 to June 25, 1984
- RCAF Station Val d'Or, Quebec, November 30, 1965, to March 31, 1975
- RCAF No. 447 Bomarc Squadron, La Macaza, Quebec, December 1963 to September 1972
- RCAF No. 446 Bomarc Squadron, North Bay, Ontario, December 1963 to September 1972
- No. 1 Fighter Wing, Lahr, West Germany, Early 1964 to December 31, 1971[17]
- No. 3 Fighter Wing, Zwibrucken, West Germany, Early 1964 to December 31, 1971
- No. 4 Fighter Wing, Baden-Sollingen, West Germany, Early 1964 to December 31, 1971

Towards the later part of the commitment, the Special Armament Storage (SAS) facilities at RCAF Station Chatham were closed. To compensate for this, a squadron capable of using the weapons would deploy from RCAF Station Chatham to RCAF Station Bagotville to be armed.

G8 Pumper undergoing an annual service test at RCAF Station Val d'Or in 1967.

---

[17] The dates for the European units are believed to be accurate, but could not be confirmed.

Physical Security

Special armament storage facilities were required at every nuclear-capable unit. Warheads used in Bomarc missiles were secured in the nose of the actual missile. Additionally, every silo had intrusion alarms installed. Safe storage of weapons used to arm aircraft required reinforced concrete structures called igloos. The igloos were located within a compound surrounded by security fencing and with armed guards positioned in towers spaced along its perimeter. Entry into the compound through a double-gated pen bordered by security fencing called a sally port. Constant patrols by armed military police were required and high-security measures that were meticulously maintained around these weapons had a direct impact on the conduct of fire responses.

Operations—Restructuring to Meet the Need

Introduction of nuclear weapons had profound and far-reaching effects on firefighting operations. Security clearances had to be upgraded, emergency response tactics changed, and controlling fires with a nuclear involvement practised. In addition, substantial increases in both personnel and equipment were required. The increases in fire department personnel and responsibilities ushered in changes in fire department rank structure as the fire departments on all nuclear-capable airfields had to be commanded by commissioned officers, where formerly they had been headed by a warrant officer first class or warrant officer second class. The exception was the Bomarc squadrons, where an upgrade in their chief's positions did not change from sergeant to flight sergeant until much later.

New Responsibilities—New Language

The learning process for the firefighter employed at these nuclear-capable sites was complicated and involved frequent testing by fire department management, by base/station authorities and by headquarters formations. Firefighters employed at nuclear-capable sites had to have a security clearance to at least secret, successfully complete the required written nuclear safety test at set intervals, master new tactics to successfully fight fires with high explosives complications, and become comfortable with all the jargon concerning security and safety procedures related to nuclear warheads. Terms such as SALLY PORT, BROKEN ARROW, BENT SPEAR, NO LONE ZONE,[18] and a myriad of closely guarded idioms entered the firefighters' vocabulary.

For the senior members of the department, attendance at an On-Scene Commanders (OSCAR) course was mandatory. Successful completion of this course enhanced the ability of graduates to take control of a nuclear accident site. When carrying out on-scene control duties they became the commander's eyes on the scene. Close contact by radio and constant updated situation reports (SITREPs) to the command post became of paramount significance. Shouldering the responsibility of the position of OSCAR was a true test of an individual's grasp of intricacies of command and control. Adopting the OSCAR title meant accepting responsibilities that included co-ordination of all on-site response agencies, personnel, and equipment and ensuring the transmission of any pertinent information to a command post, not to mention putting out the fire! To be accurate, the fire chief typically was expected to handle the OSCAR role until operations on the fire ground ceased and would then handover control to a designated OSCAR who was usually an air traffic control officer.

The rationale for the handover was not evident during an exercise when "time compression" was used extensively for the purpose of allowing all elements of the responding forces to demonstrate their capabilities, while at the same time being able to complete the evaluation within three to four hours. To achieve this,

---

18   See Appendix B for definitions.

at various points during the exercise, the evaluators would declare that a given amount of time had elapsed. Depending on the circumstances, this might be several days, as could well be the case with a real accident. Consequently, there was a legitimate need to demonstrate that there were officers qualified to perform the OSCAR role in addition to the fire chief.

Firefighting and the Nuclear-Hazard Considerations
The techniques used for fighting fires involving nuclear weapons are fundamentally the same as those used in combatting explosives fires but with the added potential of radiological hazards. That was how armament experts described the hazards to the firefighters! Firefighting efforts would usually concentrate on saving lives and, if possible, stabilizing the situation. Should the fire chief decide that the firefighters were unable to control the fire, he would declare a "simulated" BROKEN ARROW and quickly sound the withdrawal signal using the fire vehicles siren. They would then withdraw to a previously determined point at least 1,200 feet / 365 m upwind from the incident. Other first-responders would retire to a withdrawal sector not less than 2,000 feet / 610 m from the scene.

It was anticipated that small chunks of high-explosive material used in these weapons would be scattered and could be detonated by merely stepping on or kicking them. This presented a particularly dangerous hazard, given that there was likely to be foam blanket covering the ground that could conceal these fragments of explosive. Firefighting operations under these conditions could be extremely hazardous, and the only answer lay in developing safe drills and constant practice. This approach was instituted by every fire department with nuclear weapons responsibilities.

Standard of Performance
Accepting, as we must, that the Fire Service is organized for and expected to provide high-quality performance under the urgency created by emergency circumstances, for the fire departments at units conducting operations in direct support of either NATO or NORAD, the performance bar was set high. This was particularly so where nuclear munitions were concerned. Fire department performance was the subject of frequent evaluation through locally conducted operations exercises, as well as by teams from Command Headquarters and NDHQ. These teams, in most cases, were made up of both Canadian and United States personnel in Canada and included personnel from other NATO member countries in Europe. The evaluations by Operations Command formations were called Tactical Evaluations (TAC EVAL) or Operations Evaluation (OP EVAL), with the latter being the more descriptive term. Where NDHQ was involved, the evaluations were called Capability Inspections (CI).

The TAC EVAL / OP EVAL purpose and content differed starkly from the CI. A CI was conducted under peacetime conditions for the purpose of assessing the capability of the unit to store, maintain and handle nuclear munitions and to deal efficiently with an accident, called a BROKEN ARROW, involving a nuclear weapon. BROKEN ARROW exercises were, of necessity, complex. In sharp contrast, TAC EVALS / OP EVALS were keyed to simulated wartime conditions, rigidly imposed. For the fire department, it was more or less non-stop activity. Structural firefighting, aircraft rescue and firefighting, aircrew extraction, retrieving and resetting aircraft arrestor barriers, operating a fallout shelter, and operating under conditions of radioactive fallout made for a hectic three days or so.

The exercises, although taxing to be sure, could also be quite boring. The adrenalin rush associated with emergency response and action on the fire ground soon dissipated when those actions were completed. Physically securing the accident site, determining the extent of any radioactive contamination, and so on, while highly important, was tedious and slow moving.

The USAF maintained a detachment at all Canadian nuclear-capable bases. Personnel making up these detachments blended seamlessly with RCAF members to form the various specialized teams needed to minimize the severity of the accident and to assess the amount of damage suffered. Almost invariably the fire department was the lead element, going directly to the accident site to initiate damage control measures. During the early stages, the fire department would have overall command.

A failure on the part of the fire department during any of the evaluations or inspections could result in the entire base/station being deemed to have failed. This state of affairs would be brought to the attention of Canadian and United States headquarters formations. This would be followed in short order by a revaluation or re-inspection. There was no place to hide. Fortunately, there was rarely any need to try.

Turning things around a bit, there was a tremendous feeling of satisfaction in looking back at one of the exercises described here with the knowledge that all requirements were either met or markedly exceeded.

Far-Reaching Effects on the Fire Service
At any given point during these years, something in the order of 300 firefighters would be serving in this environment. It was unsurpassed as a vehicle for developing a high standard of professionalism within the Fire Service, most particularly in the exercise of command and control of response forces that generally included various elements of organizations external to the fire department under conditions of emergency.

There was an extraordinary amount of training, preparation and evaluation conducted on every nuclear-capable installation. Firefighters would form an element of every exercise scenario, and this effectively kept them in a constant state of readiness.

Demands for operational excellence made day-to-day activities much more strenuous for the average firefighter. On the positive side, the intensified training preparations and increased scope of operations surrounding the fire department activities had a positive and far-reaching effect on individual firefighters, transforming them into ever more proficient tradesmen.

Training to Meet the Demands
In 1970 in Lahr, development of a physical fitness program was undertaken to provide a measure of assurance that firefighters could safely and effectively meet the mental and physical demands the environment imposed. One hour was set aside during each shift worked for physical fitness training, where participation was mandatory and evaluations were conducted annually. The program caught the interest of *Der Kanadier*, the

A fire crew in training. WO Ian "Moe" Morrison, Cpl Norval "Benji" Benjaminson, Cpl Peter McCabe, Cpl Randy Coates, Cpl Don Shlutz, Cpl Serge "Beep" Boucher, Cpl Arnie Kaland, Cpl Vance Hull, Cpl Vern Asselstein, Cpl Glen McQueen, Capt Lorne MacLean, Cpl Lyons (legs only showing) and Cpl Chevrier.

base newspaper which published a photograph of a training session with the following comment.

Local Firefighters Keep Fit

The Lahr firefighters have been given "running commentaries" of the base scenery under the youthful guidance of Capt Lorne MacLean, base fire chief. Statistics from the fall 1972 testing period for the 1½ mile fitness run in CFE show an impressive record for the 33 firefighters in Lahr. 12 received the "good" category, and the remaining 21 scored in the "excellent" category. The base fire chief, Capt L. MacLean, lead man in the photo has earned the right to be out in front since he ran his test in a time of 8 minutes, 6 seconds, the best recorded for any age group.

Testing was done based on the CF standards rather than those of either firefighter-specific physical fitness programs (FFSPFP) which were not yet developed.

The Nuclear-Training Dividend

Fortunately, these nuclear firefighting skills were never used on an actual incident, and Canada ended its association with nuclear weapons without a single accident or incident of any gravity. The uncommon safety precautions and vigilance applied to the care, handling and maintenance of these weapons paid big dividends for the Canadian Forces by avoiding accidents or incidents of any gravity.

MISCELLANEOUS COMMITMENTS
Aircraft Arrestor Engines

The operation of the aircraft arrestor equipment fell to the firefighter. This was natural fallout, because an aircraft that engaged the cable was usually in some kind of trouble and was, therefore, already the focus of the firefighters' attention. Also, the fact that the fire department was normally the first on the scene and had the manpower, equipment and aircraft familiarity to deal appropriately with the situation made the fire department response and arrestor gear operations inseparable.

The first arrestor system inherited by the fire department was the "linear hydraulic." This system incorporated tubes buried underground and parallel to the runway. It functioned through the principle of hydraulic resistance generated by the engaged pendant cable pulling cable-mounted, tapered, cylindrical wedges through the glycol-filled tubes that were similar to water mains. Resetting the system after an engagement was quite a complex operation that required about 20 to 30 minutes to complete. With 1,500 ft / 460 m per side of nylon rope set in boxes arranged to run out smoothly at high speed, the linear hydraulic arrestor barrier was the most labour-intensive system firefighters would encounter, and it was always a challenge to get the gear reset before the next aircraft in was on final approach.

Another system firefighters had to learn to operate was the rotary hydraulic system that used a retarding mechanism in the form of a paddle within a chamber filled with a water/glycol mixture that slowed and eventually stopped the aircraft. The generated retarding force was proportional to the impact force of the aircraft. There was also a rotary friction system that employed a friction brake, some of which were adapted from the USAF B-52 aircraft that proved to be a very efficient piece of equipment that was easy to operate.

Chain barriers were also occasionally used, but generally to backup other primary systems, as they clumsy and inefficient. This system comprised a large pendant chain with huge links weighing approximately 18 kg / 40 lb each. The system had three main components that included two sections of chain approximately 1,000 ft / 300 m in length, a pendant cable and a tensioning mechanism to keep the pendant cable taut across the runway. The chains were positioned along each side of the runway, and when an aircraft engaged the barrier it would begin by dragging one link of chain from either side of the runway. As the aircraft moved along, it would drag an ever-increasing number of links as they folded back over themselves. The stopping principle was based on the aircraft engaging the pendant cable and being slowed/stopped by the friction caused by dragging the chains.

Operating the arrestor gear had its inherent dangers for the unwary firefighter. The cable tension made the area within the "V" of the cable and immediately outside a very dangerous place until the tension could be released. Aircraft exhaust and intakes presented a significant hazard in their own right, and the individuals involved in removing aircrew or disengaging the cable had to maintain a set distance from these areas. Aircraft that engaged the arrestor cable would occasionally be carrying weapons, a hazard of great potential to the responding crew. There was also the aircraft itself that provided the additional hazards to the crew with ejection seats and the possibility of overheated wheels and tires, so a response to an aircraft engaging the arrestor system was a situation that could not be taken lightly. To underscore this, one firefighter lost his life at CFB Goose Bay, Labrador, in 1990 when an arrestor cable under high tension released suddenly. Under the force of its violent reaction, it struck the firefighter and inflicted fatal injuries.

Firefighters have also to deal with the mobile-arrestor systems used mostly on fighter aircraft deployments in Canada's far north or deployed to an airfield without an in-place arrestor system. These deployable systems are generally of the rotary friction type.

UN/NATO/NORAD Missions

Members of the Fire Service have repeatedly participated in missions under the auspices of the UN, NATO and NORAD, in direct support of military operations. As a means of capturing and illustrating these important assignments, the appropriate records for a 22-year period beginning in 1982 were examined and summarized. The results are shown in Appendix "G".

SUMMARY

Escalation—Steady State—Decline

A period of growth occurred during World War II which, in turn, was followed by sharp reductions after war's end. Beginning in 1950 and continuing for more than a decade, expansion rivalled or perhaps exceeded the build-up that took place during World War II. As history unfolded, Canada's participation in NATO and NORAD was to have unprecedented impact on the Fire Service of the Air Force. By the end of the 1950s, Canadian Colours would fly over five airports and one headquarters formation in Western Europe, and there would be approximately 40 radar stations beginning operations in Canada. At some points during this period, the Air Force had upwards of 70 fire departments. With the threat inherent in the Cold War continuing unabated, strong counterbalancing measures were needed. Most notable for Canada was obtaining, storing, handling and preparing for using nuclear munitions at several sites both in Canada and Western Europe. As has been shown earlier, the implications this had for the Fire Service was especially significant.

Beginning with the 1960s, conditions of stability prevailed. For much of this decade there were 60 or more fire departments. For the 1970s and 80s, with the exception of some minor losses resulting from the closure of a few sites, this steady state was maintained. The decline of the Fire Service began to take hold in the 1990s, coincident with the seemingly inexorable reduction in the resources of Canada's military services. The Fire Service, of course, declined in lockstep.

The Comox Fire Hall opened in 1974. This photo was taken in 1990.

TALES for the TELLING

Overview
During the process of researching and compiling matters relating to the Fire Service respecting Canada's commitments to NATO, many of which were quite awesome, it became evident that although the overall picture had sufficient scope and detail, there remained a gap. The factual matters had been dealt with; however, this was not so regarding some of the more embarrassing or humorous incidents that have occurred or the flavour of the experiences absorbed by many firefighters. A sampling of these incidents and personal experiences was, therefore, sought and obtained.

Where personal experiences are concerned the 1950s, most particularly during the early to mid years when the conditions at the European units were exceptionally humble, provided fertile ground to explore for tales. This comes across quite clearly in the experiences described in "Grostenquin in the Raw" and in "Marvellous Marville." Pleasingly evident is the esprit de corps and the humour the trying conditions seemed to foster. In another vein, RCAF Station Rockliffe provided irrefutable proof that Murphy's Law is relentlessly pervasive.

Embarrassment Personified
Yes, it not only can happen, it did. In 1946, the fire hall at RCAF Station, Rockcliffe, Ontario, burned to the ground. The fire also destroyed a G10 pumper and a G15 crash truck, which was the entire inventory of apparatus at the time. The fire had advanced so rapidly that the duty crew were fortunate to escape with their lives through the dormitory windows. The RCAF Station Uplands, Ontario, Fire Department was called but due to the advanced stage of the fire, coupled with a response distance of approximately 12 miles / 19 km, it was decided that it would unwise to leave Uplands without fire protection for what would almost assuredly be a futile effort.

Although there have been fires in other fire halls, it is believed that this was the only fire hall to be destroyed by fire. Most embarrassing, to be sure, and made the more so by the fact that the most probable cause of the fire was determined to be a smoothing iron left turned on and unattended on a wooden ironing board. At the time, single firefighters lived in the fire hall, hence the need for laundry facilities.

The remains of the Rockcliffe Fire Hall.

Over the Hill
In the early 1950s an event occurred at RCAF Station Rockcliffe that seems to indicate that Murphy's Law was being applied with excessive zeal at this station. Having had its fire hall destroyed by fire in 1946, another embarrassment was in the works. It played out like this. One night shortly after midnight a member of the duty fire crew took the G10 pumper to drive up a hill to do a fire-safety inspection of the officers' mess. The mess was located on a hilltop with a commanding view of the airfield, the Ottawa River and portions of the Gatineau region of Quebec. On arrival at the mess, the crewman parked the pumper and went inside to complete the inspection. When he came back outside,

the pumper was gone. Apparently, he had not set the parking brake firmly enough, and it rolled forward and dropped over an almost vertical rock-faced cliff—hence Rockcliffe—and suffered severe damage. The firefighter was found guilty of negligence and was compulsorily re-mustered to heavy-equipment operator.

Grostenquin in the Raw[19]

From the Memoirs of Chief Warrant Officer Jim Lockhart (1951–84), as recalled in 2004.

> It all began during the summer of 1952. At that time there were a lot of rumours circulating in RCAF Station St-Jean, Quebec, where I was serving, about the opening of an Air Force Station in Europe and that volunteers to serve there were being sought. In response to this, some of my co-workers had volunteered with no results. I had only been in St-Jean for six months and was enjoying myself, so I was not really interested in Europe at the time. However, I was being ridiculed by my peers for being a bit of a chicken. Perhaps as a result of this, sometime in September I went over to the Station Orderly Room to inquire of the procedure for volunteering for an overseas posting. The clerk took my name, number and so on and asked me to have a seat. In about five minutes she returned with a message (telegram) in the hand, and needless to say I was surprised to be informed I was posted to No. 2 Fighter Wing, Grostenquin, France, on October 16, 1952 and that I had better start the process for an overseas posting. One can imagine the surprise I was met with upon returning to the fire hall with this hot news.
>
> Time seemed to spin forward with uncommon speed. Medical examinations, immunization, pre-embarkation leave were all completed and on October 16, 1952, I, along with 97 other Airmen boarded the ship *Columbia* in Montreal for a six-day journey to the Port of Cherbourg, France, where we were met by a guide who took us to Paris for the weekend. On Monday we took the train to Foulquemont and arrived that night.
>
> We were met by military buses that took us to No. 2 Fighter Wing, Grostenquin. I remember getting off the bus at the barracks late in the evening. It was quite dark and it seemed there was mud everywhere. It was so bad the first people that got off the bus tried to get back on. Members of the advance party were in the back ground with the usual "Welcome to Grostenquin" chants and signs and helped us get settled into our barracks.
>
> The next morning was an eye-opener. The streets had not been paved and there was plenty of mud. Utility poles lined the streets with wires and pipes on them and the buildings were of the prefabricated type. It was a sorrowful looking sight and I thought to myself, my tour of two years would seem like a long time. After breakfast we had an in-briefing by the commanding officer and the wing warrant officer. A couple of things I remember from the briefing was, we were seven minutes flying time from the nearest Soviet Union forces, and we would be "confined to barracks" on May 1 which, in reality, meant we could not leave Wing

---

19   Memoirs of Chief Warrant Officer Jim Lockhart (1951–84).

Property. This was done in deference to the activities May Day celebrations brought forth in the Soviet Union which were built largely on a display of military might. These activities were an effective means of impressing on all concerned the tensions that would prevail for Canada's European Units and those of other NATO nations for some three decades to come. There were also some general instructions on how to avoid doing anything to upset the local residents.

The situation was a bit unique in that it was an all-male contingent. There were no airwomen on the Wing. As well, none of the married men had their wives and children.

Next on list was to go to the supply section and get ourselves a pair of hip-wader rubber boots to protect our clothing from the ever-present mud. Following that, I reported to the fire department sometime around noon, met the staff and must have gotten some protective clothing because I spent the afternoon on the airfield with another firefighter on the G13. That afternoon I attended my first aircraft crash. I don't remember who I was with, and if I remember correctly, the F86 Sabre aircraft that was doing landings and take-offs when it lost thrust due to a "flamed-out." As a result, it went through the fence at the end of the runway. The pilot was OK and I put the fire out with a 30-lb dry chemical.

The more I look back at how things were at Number 2 FW, the better I understand why things were as they were. If I remember correctly there was a staff of 13 (some would say, a bad omen to start with): one sergeant, two corporals and the rest were aircraftsman (AC) and leading aircraftsman (LAC). I was an AC with just ten months in a Fire Department after having completed basic trade training at the fire school at Aylmer, Ontario, and had never seen an airfield. Given the fact that most of the men were, like me, young, and with little experience, the chief and the deputy chief had their hands full. They were responsible for crash protection for three squadrons of F86 Sabre aircraft that were being flown by young pilots. They also had to provide fire protection for buildings and facilities that were either in place or being built. It was the only time I ever did fire patrols in the daytime.

All the buildings had to be scaled for fire extinguishers and they had to be installed. In addition to these and other requirements, the ACs had to be prepared for the semi-annual trade-board exams and general trade advancement. Needless to say, candidates wrote the trade-boards as if predestined to fail. The questions or maybe the answers seemed to be tricky and the only publications or study material that I can recall was *Trade Précis Number 13* and a policy manual titled *Canadian Air Publication Number 123*. It was highly questionable how up-to-date either document was. You were on your own; there was no help from the more senior people.

A couple of personal things happened that are indelibly etched in my memory. The first was being selected to the honour guard for the Queen's Coronation in England. There was, however, this other lad who had served in the Army, had a number of medals, and was anxious to go, but was bypassed, so I let him take my place. Something I have, from time to time, looked back on with regret. The second happening occurred on Christmas Eve, 1952. It went something like this. In the early part of the evening I was lying in my bunk reading or just daydreaming when someone banged in my door and said, "It is Christmas Eve, do you want a beer?" I made the mistake of replying in the affirmative. From this point onwards things moved along rather quickly and within a short time there was about seven cases of beer in my room. Apparently some beer company sent over five thousand bottles of beer for the troops for Christmas and New Year's. Since half the Wing personnel had gone to England for Christmas, the ones remained had the beer.

Power failures were quite frequent. We were told the Canadian and French governments couldn't decide who was supposed to pay the power bills. As a result we often had to eat our meals cold and with candles or some of us were fortunate enough to have lanterns. Even though the conditions bordered on the atrocious, the morale was quite good. As a matter of fact, it was so good that sometime in September Flight Sergeant Maggs from the Fire Marshal's Office in Metz, France, asked for (if you can believe this) two experienced firefighters to volunteer to go to beautiful No. 3 Wing in Zwiebrucken (Two Bridges),

West Germany. No one volunteered so he picked Aircraftsman Hebbard and me. I left No. 2 Fighter Wing in October 1953.

There is one other thing I should perhaps mention. I don't know if it was when I arrived or later but I remember there was a cut-down 45 G / 200 L drum in the bays of the Fire Hall where we used to burn scraps of wood to keep warm.

"Marvellous" Marville[20]

From the Memoirs of Flying Officer Doug Stevenson (1946–63), as recalled in 2004:

> I was transferred from RCAF Station North Luffenham, England, to what was to be the new home of No. 1 Fighter Wing in Marville, France, arriving there on Monday, December 6, 1954, as part of an Advance Party. I had just driven up from No. 1 Air Division Headquarters in Metz, France, having stayed overnight at the home of Flight Lieutenant Archie Graham, the Command Fire Marshal (CFM). I drove onto the Wing and parked my 1954 Morris Minor (my very first brand new car) in front of the guardhouse and stepped out, all pressed and polished, into an inch of sloppy mud, the first of much more to come.
>
> The roads in the immediate area of the guardhouse had in fact been swept appropriately in preparation for the arrival of a VIP; no, not for Corporal Doug Stevenson, but for the Air Officer Commanding (AOC) of No. 1 Air Division, in the person of Air Vice Marshal Hugh Campbell. As it turned out, I had arrived

---

[20] Memoirs of Flying Officer Doug Stevenson (1946–63).

just prior to a ribbon-cutting ceremony of the first building to be officially handed over to the Air Force by the Contractor. The AOC and his party arrived via DC3 Dakota aircraft about the same time as I did. Staff cars had been driven up from Metz to ferry the VIP party from the aircraft to the chosen barrack block, then back to the aircraft. The barrack block was located kitty-corner from the Guard House, so I was able to watch the ceremony whilst standing in the mud.

Once the official stuff was over and the VIPs had left, I reported to my new boss, Flying Officer John Lepage, the Construction Engineering Officer. He had been at the site for over a year, overseeing the construction of the infrastructure needed to support Wing operations. He explained that my job was to nose around and learn everything I could about everything that I thought might be important from a Fire Department point of view. As well as Flying Officer Lepage, there were two corporals already there, making me the fourth member of the Air Force to be taken on strength at Marville.

The guardhouse was one of the few buildings which had been completed and was to become my home for the next couple of months. Within the next week or so several military police corporals and a mobile equipment driver arrived and we were all housed in the Guard House. Beds were moved into the offices and the four cells. The Air Force personnel occupied the offices, while four French gendarmes moved into the cells. The gendarmes were there to police the contractor's labour force, which was comprised of 100 or so Algerians, who all lived in a slum-like conglomeration of shacks, just off the site.

At first, the water was unfit to drink and, from time to time, I had to clean my teeth with wine. This situation prevailed for just a few days after which a water-tank trailer brought in potable water from ADHQ in Metz and parked outside our backdoor. At the same time, a supply of canned food, etc. was brought in from the post exchange (PX) at ADHQ in Metz. We fed ourselves breakfast and lunch from this food stash, but dinner was different. Some clear thinking and seemingly smooth-talking person somehow negotiated with a local hotel, about 15 km / 9 miles distant, to serve dinner each night to the eight or ten of us who lived in the guardhouse.

For me, this was to be an experience which I still remember fondly. We habitually arrived at the Hotel de Commerce in Stenay each evening at 1900 hours via the "Prevost" military bus and were served the *table d' hôtel*. This was my introduction to French cuisine. We would start off with an aperitif followed by the most delicious food I had ever tasted, accompanied, as it was, by copious amounts of *vin ordinaire*. These meals usually lasted from two to three hours, ending with everyone, except the driver of course, in a very mellow mood. We would then all pile onto the bus for the half-hour trip back to the Wing to flop into bed and sleep like babies.

With regard to the fire hall itself, one thing sticks out in my mind. It was built on a slight hill, the grade was level with the rear bay doors, but the apparatus room floor was about four feet above grade at the front. Two sloping concrete walls had been poured, one on each side of the overhead doors. One day several trucks arrived and dumped tons of stone about the size of watermelons between the walls. Two of the Algerian labourers arrived equipped with long-handled shovels and nothing else and, incredibly, within about three days, they had built a smooth ramp from the floor to the road level—very impressive.

The supply officer was Flight Lieutenant Jim West who, years later, was promoted to group captain and served as commanding officer of RCAF Station Rockcliffe. This officer was really good at his job. Whatever we needed, he got for us in record time. He even travelled to the Post Exchange at No. 2 (F) Wing, Grostenquin, France, on the weekends to fill our needs. He really was a supply officer—in every respect. All in all, although very trying at times, it was, indeed, a great experience.

Where the Dust Settled in No. 1 Air Division Headquarters Fire Department, Metz

There is a humorous tale about the fire department serving this headquarters. It seems that during an official visit to the Headquarters Fire Department, the fire marshal became somewhat disenchanted with the standard of cleanliness of the fire hall and the fire apparatus. He proceeded to let the fire chief know of his displeasure in no uncertain terms. In fact, he was so annoyed by the situation that he surreptitiously wrote his initials in the dust on the pumper! Weeks later, he returned on an apparently unscheduled visit only to find his initials still gracing the dust on the pumper! The fallout of this return visit is not known.

CHAPTER V

PERSONNEL

Overview

There are many subjects recorded elsewhere in this book that could well have been placed in this chapter. This is so since there are but few subjects that do not embody a personnel component. Conversely, these selfsame subjects had other components that, on some occasions, were more dominant. In many cases it was simply a matter of choice as to where a specific subject area should be recorded. Typical of this was the various issues associated with unification of the armed forces, which had significant effect on both individual firefighters and on the organizational structure of the Fire Service. The seeming permanency of the organizational changes resulted in the subject being addressed in Chapter I.

The physical fitness of firefighters is a subject that is almost exclusively personnel-oriented. For something like the past quarter of a century it has been a subject of much interest. With the aim of creating conditions that provide a measure of assurance that firefighters can perform their sometimes demanding tasks effectively and safely, physical fitness programs, oddly enough, were sometimes quite divisive.

Firefighters have been the recipient of many honours and awards as is reflected in this chapter, while the extensive involvement of firefighters in military operations is summarized in Appendix "H".

Also, irrespective of how dedicated and serious-minded any of us may be in the Fire Service, the camaraderie that is its linchpin will inevitably show through. This is reflected in "Burnishing the Ties That Bind," as described later in this chapter.

Conditions of Employment

Over the years, the hours of work for firefighters have undergone major changes from 84 hours per week to 56 hours per week and finally to 42 hours per week. Each reduction in the number of hours worked increased the number of crews required to provide uninterrupted service 24 hours per day, seven days per week. With the 84-hour week, the coverage was achieved using two crews working on a 24 hours on duty, 24 hours off duty shift arrangement. This changed in the late 1950s / early 1960s, when the 56-hour work week was adopted in which three crews were used with a shift arrangement of three ten hour day shifts, three 14-hour night shifts followed by three days off. With the implementation of the 42-hour work week in the mid 1970s, the shift arrangement remained much the same, except that four crews were required, and it provided for six days off instead of three. There were no increases to establishments to offset these changes. Great care had to be exercised respecting deployment of staff in order to provide suitably qualified supervisors/commanders for the additional crews.

Advancing with the Times

Beginning in the 1950s, and continuing for three or four decades, a series of changes and events that had significant impact on the Fire Service took place. It was an era where the Fire Service found itself on the front line of technological advances in firefighting techniques, equipment, materials and physical-fitness training, as well as an ever expanding role. During this exciting and progressive era, the individual firefighter realized levels in training, personal fitness and a scope of operational activities that his predecessors could hardly have imagined.

Changes—Fire Department Rank Structure

The military rank structure in fire departments and terminology used to identify the various positions also underwent changes. Amongst other things, designation for the position of crew chief was changed to platoon chief. At some locations, the rank established for the position was increased from sergeant to warrant officer, a rank increase that positively affected firefighters by allowing greater scope for advancement. In the main, these changes came into being in the late 1950s to early '60s coincident with the need to increase the level of fire protection to counteract the critical increase in risk associated with the storage and handling of nuclear munitions. This eventually became policy for all locations requiring Aircraft Rescue and Firefighting Services (ARFF) of Category 6 and above.

PHYSICAL FITNESS

Background

Firefighting is a hazardous profession that periodically places intense psychological stress and physically taxing demands on its members. The issue of poor and unsafe states of physical fitness of military and civilian firefighters was first addressed at a conference of Canadian Forces Fire Marshals in early 1977. Records show that 40% to 50% of firefighter deaths across North America are from heart attacks. DND has long recognized this, and for many decades has had a health program involving both medical evaluations and physical fitness standards.

The senior fire protection officers of this era were the prime proponents of this program and expended much energy promoting its development. Although this program met considerable resistance in many quarters, finally standards were written, and the program obtained general acceptance. This set of standards also became widely recognized as a mandatory requirement to become a firefighter in municipal fire departments.

First FFSPFP

Shortly after that conference, FFSPFP was developed, based on the National Fire Protection Association

(NFPA) book titled *Physical Fitness and the Fire Service*.[1] The program was selected and adopted in 1978 and consisted of the same evolutions identified as minimum entrance requirements in the NPFA Firefighter Professional Qualifications Standard.[2]

Training Practices

At that time, it was mutually agreed by fire marshals that one hour for group participation during each shift worked was to be scheduled during working hours, and that job descriptions and terms of references were to be amended to include physical fitness participation as a duty during this period. Physical fitness training by professionals was also available if requested.

Evaluation Standard

The test was comprised of six items as follows:

1.5 Mile Run

Run 1.5 miles / 2.4 kms within the time frame for his/her age group. The purpose of this test was to measure the condition of the lungs, heart and vascular system, and the standard represented what was considered by most authorities to represent a minimum requirement for overall physical fitness in an adult.

Push-Ups

Complete the number of push ups established for his/her age group. This test was designed to simulate the firefighter's need to push on the fire ground as may be required in the use of pike poles, battering rams, and other equipment.

Sit-Ups

Perform the number of bent-knee sit-ups established for his/her age group within ninety seconds. The purpose of this test was to determine the firefighter's ability to resist hernia injuries and to provide an indication of the strength of body support muscles.

Pull-Ups or Chin-Ups

Complete a total of five consecutive pull-ups or chin-ups. This test was designed to simulate the firefighter's ability to lift or pull equipment, hoses, ropes and other tools on the fire ground.

Balance

Walk the length of a secured balance beam measuring 20 ft / 6 m in length by 3.5 in / 9 cm in width, while carrying a length of fire hose weighing 20 lb / 9 kg without falling off or stepping off the beam. The beam walk was designed to demonstrate the individual's ability to maintain vertical balance of the body when the footing may not be sound.

Man Carry

Lift 125 lb / 57 kg from the floor and carry the weight 100 ft / 30 m without stopping. This test was designed to assess a firefighter's ability to carry a victim in an emergency situation.

Mandatory Rest Periods

A minimum of five minutes and a maximum of 10 minutes were allocated as rest periods between completion of the 1.5

---

1 A-CE-F50-002/PT-001. *Physical Fitness and the Fire Service.* National Fire Protection Association.
2 NFPA 1001, Standard for Firefighter Professional Qualifications.

mile / 2.4 km run and the start of the push ups. In addition, a minimum of one minute and a maximum of five minutes were allocated as rest periods between the subsequent test items.

Administrative Hurdles

Expressions of disapproval of this program and the lack of active support at the fire department management levels were raised by firefighters. Questions were also raised with regards to the compulsory medical examinations that were to be administered prior to the first annual test.

Two particular problems were identified. First, the timings of the National Health and Welfare medicals did not coincide with the timings of the fitness tests, and second, the revised Occupational Health Evaluation Standard required annual medicals only for firefighters over the age of 45 years. Thus, the program was delayed until November 1980, when medical authorities concurred that the program could proceed, providing certain conditions were met. Based on these conditions, a Physical Fitness Testing Directive was issued by NDHQ on January 30, 1981. Following are the abbreviated highlights of that directive:

- all firefighters (regardless of rank or employment) were subject to physical-fitness evaluation testing;
- the annual test period was set from May 31 to June 30 each year, with the first test to be completed in 1981;
- the evaluation tests were to be administered by military Physical Education and Recreation Instructors (PERI), and the test results were to be compiled by fire chiefs and forwarded to the CFFM through the appropriate CFM each year;
- in case of failure and/or excusals, a follow-up test and reports were to be submitted on a quarterly basis until the matter was resolved;
- firefighters who had been actively participating in the program and who had undergone a Category III medical examination and obtained a Class A certification (fit for work as annotated on Health and Welfare Canada form 365) within two years of the physical fitness evaluation test could undergo testing without requiring further medical examination; and
- all other firefighters were required to undergo a Category III medical examination and be declared "fit for work" before attempting the physical fitness evaluation test.

National Policy

The physical-fitness standards and supporting policy for conducting of the physical-fitness test for both military and civilian firefighters was promulgated in Canadian Forces Administrative Order (CFAO) 50-23[3] and Civilian Personnel Administrative Order (CPAO) 9.21.[4]

Complaints—First FFSPFP

In 1981, complaints were formally submitted by civilian firefighters to the Anti-Discrimination Directorate of the Public Service Commission (PSC) concerning the fitness standards established for the 1.5 mile / 2.4 km run test. The complaints were based on the fact that the DND firefighter standards for aerobic fitness were higher for both males and females in all age groups than the internationally recognized Cooper standards (American Health Association) for overall aerobic fitness. The PSC of Canada, Appeals and Investigations Branch, formally investigated these allegations and requested the Directorate of Physical Education and Recreation (DPERA) and the CFFM to address the complaints.[5]

---

3   Fire Protection Personnel—Physical Fitness.
4   Firefighter Group Physical Fitness.
5   Ref: 81-DND-213 dated December 22, 1981. Public Service Commission of Canada, Appeals and Investigations Branch.

Complaints Answered

On February 9, 1982, DPERA[6] provided a written response to the Investigation/Conciliation Review Officer of the Anti-Discrimination Directorate (ADD) with respect to the established standards for the 1.5 mile / 2.4 km run. Following are abbreviated highlights of the written response:

- The Cooper standards (1977) are guidelines developed by Dr. Kenneth Cooper for a general population and do not necessarily reflect the needs of any specific group of personnel in specialty trades, such as firefighter.
- The standards established by the CFFM in consultation with DPERA are neither at the superior level of the CF Standards (CFAO 50-1) nor are they at the minimum acceptable (good) level.
- It is not uncommon for many CF specialty trades, because of the role they are expected to perform, to be more stringent than the norm for physical fitness standards.
- DPERA supports the 1.5 mile / 2.4 km run test standards established by CFFM for all DND firefighters.
- Firefighters are a specialty category and being fit will enable them to exercise their special responsibility for protecting DND personnel and their families in the event of emergencies.

Over the next year, the PSC, Appeals and Investigations Branch, received complaints alleging that the physical fitness test administered to civilian (DND) firefighters was discriminatory on the grounds of age and sex, and unfair in that the standards being applied at that time were higher for civilians than for military firefighters.[7]

With respect to allegations of the test being discriminatory on the grounds of age and sex, it is important to document for the record that the standards for the 1.5 mile / 2.4 km run, push-ups, and sit-ups at that time were stratified by both age and gender. Thus female and older firefighters were required to meet a lower standard than their younger male counterparts. Hence, in essence, the civilian firefighters were complaining that older and female firefighters did not have to meet the same standards of physical fitness as their younger counterparts, although all firefighters were required to complete the same job. These complaints eventually formed the philosophical basis for establishing the current gender- and age-free standard.

With respect to allegations of the standards being unfair, the same standards were applied to both civilian and military firefighters except that career sanctions were only applicable to military firefighters. Therefore, if the application of the physical fitness standards was held to be unfair, the unfairness applied to military firefighters rather than civilian firefighters.

Forces-Wide Physical Fitness Program

In order to gain a better understanding of the communications between the PSC, Appeals and Investigations Branch, and DPERA during 1982–83, it is necessary to document the concurrent development and implementation of physical fitness standards for CF personnel. Prior to 1980, CF members were evaluated annually by means of a 1.5 mile / 2.4 km run, sit-ups, push-ups, and chin-ups. This annual test had age- and gender-based standards, and was derived from the work of Dr. Kenneth Cooper (1968)[8] and Astrand and Rhyming (1954).[9]

---

6 Ref: 4816-2-0 (DPERA 3) dated February 9, 1982.
7 Ref: 81-DND-214-ADD, dated November 3, 1981.
8 "A Means of Assessing Maximal Oxygen Intake." *Journal of the American Medical Association.*
9 "A Nomogram for Calculation of Aerobic Capacity (Physical Fitness) From Pulse Rate During Submaximal Work," *Journal of Applied Physiology.*

During the early years of the firefighter physical fitness program, the CF, as a whole, used a test to evaluate the physical fitness level of the general military population that included a 1.5 mile / 2.4 km run test. However, due to the number of injuries and deaths associated with the run test, in September 1980, the CF Surgeon General deemed the test as unsafe for personnel over the age of 30 and cancelled it for the general military population.

The test itself was not unsafe; the problem lay with the manner in which the test was administered, and the lack of supporting training programs to properly prepare CF members for a maximal test. Specifically, pre-screening to identify personnel for whom the test may be inappropriate (e.g. health appraisal questionnaire, measurements of resting heart rate and blood pressure) was not conducted prior to members being administered the run test. In addition, published supporting training programs were virtually non-existent, resulting in personnel being unprepared physically to exert a maximal effort during the run test.

It must be noted that although the run test was cancelled as a test for the general military population, it was still retained for specialty trades, such as search and rescue, firefighters, and physical education and recreation instructors, as well as unique units, such as the airborne regiment and parachutists.

With the cancellation of CF's physical fitness evaluation, the Chief of Defence Staff (CDS) instructed that a new approach to physical fitness evaluation and training be developed. The initial outcome of this direction was the adoption of the Department of Health and Welfare, Fitness and Amateur Sport, Canadian Standardized Test of Fitness (CSTF)[10] as the annual evaluation for CF members.

CSTF

The CSTF adopted by the CF was comprised of a multi-stage progressive step-test to assess aerobic capacity, push-ups and sit-ups to measure upper body and abdominal core muscular endurance, and hand-grip to measure upper body muscular strength. The CF physical fitness program from 1980–83 was called Project Phoenix, and there were no established physical fitness standards for CF personnel during this period; however, the CF developed unique exercise training programs in support of the evaluation. The new CF physical fitness program, consisting of the CSTF and supporting training programs, was renamed the CF Exercise Prescription (EXPRES) Plan and was approved by the Defence Management Committee (DMC) in February 1983 and was promulgated in December 1983.

DPERA and CFFM felt that the CSTF could be adopted to replace the civilian and military firefighter physical fitness test as given in CFAO 50-23 and CPAO 9-21. However, this never came to fruition, as in December 1983, the CDS directed that minimum physical fitness standards for CF members be researched and developed based on bona fide occupational requirements (BFOR) as stipulated in the Canadian Human Rights, Bona Fide Occupation Requirement Guidelines.

Research—Minimum Physical Fitness Standards for the Forces

Between 1983 and 1988, DND contracted the Queen's University Ergonomics Research Group to research and develop minimum physical fitness standards for CF personnel, based on the physical requirements of CF members performing common military tasks. Therefore, if the CSTF was to be adopted for civilian and military firefighters, the standards applied to

---

[10] Fitness and Amateur Sport *Canadian Standardized Test of Fitness (CSTF) Operations Manual*. 1st Edition. Ottawa, Ontario, 1981.

firefighters would have to be developed based on the performance of common firefighting tasks. It was further conceptualized that CF minimum physical fitness standards would be developed on the physical requirements of performing common military tasks prior to the development of unique trade fitness standards based on specific occupational requirements.

In the fall of 1989, the CF Minimum Physical Fitness Standards (MPFS) were approved and implemented as training objectives without career sanctions for all healthy military personnel.

Looking Back and Ahead
In December 1989, meetings occurred between CFFM and DPERA during which a revised physical fitness program for firefighters was discussed. CFFM identified its system of testing the physical fitness level of firefighters as inadequate for three reasons. First, career sanctions were applicable to military firefighters who failed to meet the established fitness standards, whereas there were no career sanctions applicable to civilian firefighters failing to meet the same standards of physical fitness, even though the specific job requirements were the same for both groups of firefighters. Second, the established standards had never been scientifically researched to determine if they reflected the actual demands of the job. Third, some firefighters had expressed concern that the test was not a valid or accurate measurement of their operational capabilities. Thus, CFFM formally requested DPERA to conduct appropriate research to develop an occupation specific maintenance test that would be reflective of actual job demands while ensuring that firefighters would be physically capable of carrying out their duties.[11]

New Research—Firefighter Physical Fitness Program
Based upon these discussions, CFFM requested that the Chief of Research and Development (CRAD), Directorate of Research and Development Human Performance (DRDHP) provide funding for the research and development of a new firefighter program which would be non-gender, non-biased, and task-related.[12]

The request to develop a valid physical fitness test specifically tailored for DND firefighters was supported by the CF Surgeon General Branch.[13] CRAD/DHPHP approved funding for the project to conduct research and to develop physical fitness standards for CF/DND firefighters.[14] Pending approval of a formal contract, which is a lengthy and complicated process, Queen's University was provided with initial funding by DPERA to cover the costs of conducting preliminary work which commenced on May 7, 1993, towards development of firefighter physical fitness standards.[15]

On January 6, 1994, Queen's University was awarded a contract from Supply and Services Canada, Science Branch, to research and develop a bona fide physical fitness maintenance standard for firefighters.[16]

Second FFSPFP
The second FFSPFP, which included a ten-item test circuit that represented the most demanding and representative tasks specific to firefighters, was developed for pilot testing on a group of military firefighters. The circuit was refined and evolved over a period of performance trials conducted at CFFA. The resulting circuit was presented to a panel of experts and eventually was accepted as being representative of the

---

11   4816-2-6 (CFFM) December 19, 1989.
12   4815-1-91 (CFFM) July 5, 1991.
13   4815-1-91 (DMO) August 29, 1991.
14   5595-5-6 (DPERA) January 12, 1993.
15   5595-5-6 (DPERA) October 26, 1993.
16   Supply and Services Canada. SSC File No. 003SV.W8477-3-SC02. Contract No. W8477-3-SC02/01-SV.

duties performed by firefighters. To maximize economy of time and resources, it was recommended that the physical fitness evaluation be conducted in the fire hall, with the equipment already available.

The approved circuit consisted of ten simulated firefighting tasks which were to be completed in a continuous and consecutive manner within the confines of a fire hall. Rest intervals between the individual tasks in the fitness test, consisting of walking a distance of either 15 m / 50 ft or 30 m / 100 ft were incorporated to represent the conduct of tasks at the scene of an actual fire.

The physical fitness test had to be completed with full turnout gear, within eight minutes, and include the following tasks.

One-Arm Hose Carry
Carry a rolled 30 m / 100 ft length of 65 mm / 2.5 in hose a distance of 15 m / 50 ft and return to the start point carrying the hose in the other hand. Walk 15 m / 50 ft to the next start point.

Ladder Raise
Carry a 3.5 m / 12 ft ladder a distance of 15 m / 50 ft and raise it against a wall. Walk 15 m / 50 ft to the next start point.

Hose Drag
Drag a 30 m / 100 ft length of charged 38 mm / 1.5 in hose a distance of 30 m / 100 ft. Walk 15 m / 50 ft to the next start point.

Ladder Climb
Climb a 7 m / 24 ft (ten rungs) ladder three times. Walk 15 m / 50 ft to the next start point.

High-Volume Hose Pull
Using a hand-over-hand method on a length of rope, pull a 30 m / 100 ft length of 100 mm / 4 inch of hose and a 15 m / 50 ft length of 65 mm / 2.5 inch hose that are rolled and then tied together with a rope, a distance of 15 m / 50 ft. Return to the point where the hose was originally positioned and repeat the pull. Walk 15 m / 50 ft to the next start point.

Forcible Entry
Move a tire weighing 100 kg / 220 lb a distance of 30 cm / 12 in with a 4.5 kg / 10 lb sledge hammer. Walk 15 m / 50 ft to the next start point.

Victim Drag
Drag a mannequin weighing 68 kg / 150 lb a distance of 15 m / 50 ft. Turn, and drag the mannequin another 15 m / 50 ft to the point of departure. Walk 15 m / 50 ft to the next start point.

Ladder Climb
Climb a 7 m / 24 ft (ten rungs) ladder twice. Walk 15 m / 50 ft to the next start point.

Ladder Lower and Carry
Lower the 3.5 m / 12 ft ladder used in the ladder raise exercise and carry the ladder 15 m / 50 ft Walk 15 m / 50 ft to the next start point.

Spreader Tool Carry
Carry a 36 kg / 80 lb hydraulic spreader tool a distance of 15 m / 50 ft. Turn, and carry the tool to the point of departure.

Participants in both fitness programs had access to: a qualified physical fitness trainer, fitness training equipment in every fire hall, physical fitness clothing for every firefighter, and military fitness facilities. As well, time was allocated during working hours for physical fitness training. Where the second FFSPFP was concerned, every firefighter also had access to the Strengthening the Forces Program, a health promotion program that provided guidance and assistance towards the development and practice of a healthy lifestyle.

Implementation
The second FFSPFP was officially introduced in 1997 with a four-year transition period. Although it met the BFOR described in the *Canadian Human Rights Act*, it was not well received by the civilian firefighters who saw the test as a threat to their employment as firefighters. On the other hand, military firefighters have been doing the test since 1997 with a 99.5% success rate.

Complaints—Second FFSPFP
The Union of National Defence Employees (UNDE) has not recognized the test as a BFOR and has unequivocally opposed the testing portion of the program, particularly as it might impact on job security. Despite the issuance of a directive by the Deputy Minister of National Defence to implement the test in February 2002, civilian firefighters refused to comply.

In early 2002, the Canadian Human Rights Commission (CHRC) received a complaint from two civilian firefighters alleging that subjecting DND firefighters to the DND/CF firefighter physical fitness test as a condition of hiring and ongoing employment discriminate on the basis of sex and age. This was strikingly similar to the complaints respecting the first FFSPFP. Concurrently, the same grievance was filed by all other civilian firefighters. In late 2002, two bases requested seven firefighters to take the test. They all refused, stating it was unsafe, thus adding a new dimension to the complaints. At the time of publication of this history, the issue remained unresolved.

MAINTAINING a FIRE SERVICE OFFICER CORPS

Commissioning from the Ranks Plan[17]
Commissioning from the Ranks Plan (CFRP) for the Fire Service started in 1952. Due to the lack of a senior officer to head the Fire Service, the RCAF invited an ex-wartime officer to return/re-enlist for five years as the titular head of the burgeoning RCAF Fire Service. In the early to mid 1950s, 13 officers were commissioned from the ranks. The major increase in operations and risk associated with the Canadian commitments to NATO, particularly in Europe, and to NORAD, required the number of Fire Service officers to be increased to 28.

The CFRP provides a means for a commanding officer to nominate for commissioning selected senior NCOs found to be suitable for employment as officers. The CFRP applies to NCOs who have acquired the level of military experience and personal qualities that combined gives clear evidence of the capacity to be effective in positions with a higher level of responsibility. Recently it has been opened to include outstanding master corporals.

To be eligible a member must be a Canadian citizen, have a minimum of ten years service in the CF, have at least the minimum required years of service remaining before compulsory release age (CRA), meet the minimum medical requirements for their selected military occupation code (MOC), achieve acceptable standing for officers on the Canadian Forces Aptitude Test (CFAT), and possess a Grade 12 diploma (Section V for Quebec) or an equivalency certificate.

---

17 CFAO 11-9 Commissioning from the Ranks Plan.

Special Commissioning Plan[18]

The Special Commissioning Plan (SCP) has provided a means where regular force non-commissioned members (NCM) who have the academic qualifications for enrolment as direct-entry officers may apply for commissioning. Applicants must have been identified by their superiors as possessing personal qualities and job performance indicative of officer potential, and be Canadian citizens or a landed immigrant and have not less than Trade Qualification, Level 3. Applicants must also show that they meet the minimum medical requirements for their selected MOC by having a medical profile awarded or confirmed within the 12-month period immediately prior to the competition closing date. To meet academic eligibility, an applicant must possess a baccalaureate degree, or in some cases a technical diploma that satisfies the enrolment criteria.

Special Ranks Commissioning Plan

In the mid to late 1980s, due to a shortage of officers, the Fire Service began to use the SRCP to fill vacancies. This plan differs from the CFRP in that the officers commissioned under the SRCP do not receive Canadian Military Engineer (CME) Training and have only limited employment and advancement opportunities within the Fire Service. The SRCP permitted the CF and the Fire Service to profit from the extensive skill and experience of firefighters, chief warrant officers and senior master warrant officers who are within a few years of CRA.

University Training Plan—Non Commissioned Members[19]

The University Training Plan for Non-Commissioned Members (UTPNCM) is a program whereby selected members, who do not already possess a university degree, or a degree acceptable to their applicable MOC, are sponsored for up to four years of full-time attendance at the Royal Military College (RMC) or a Canadian university. Since 2002, this plan has been open to members who have achieved the minimum rank of corporal by the closing date of the competition. In order to be eligible, they must have been identified by their superiors as possessing personal qualities and job performance indicative of officer potential.

This plan has been first and foremost an officer production program, with academic upgrading included as part of the plan. For the Fire Service, the applicable degree is fire engineering.

Applicants must be Canadian citizens or have obtained landed immigrant status and must have completed two, full, six-credit university courses or their equivalent prior to application. In this case, an equivalent is a college course that a Canadian university would recognize as meeting the requirements for a similar university course in a given program. Members must achieve acceptable standing for officers on the CFAT and meet the minimum medical requirements for their selected MOC.

SERVICE in the BROADER SENSE

Introduction

Over the years, DND Fire Service personnel, both military and civilian, have helped advance the science and technology of the Fire Service in Canada through their participation in the:

- ACFM&FC / Council of Canadian Fire Marshals and Fire Commissioners (CCFM&FC)
- CAFC
- Fire Prevention Canada (FPC)
- NFPA
- Underwriters' Laboratories of Canada (ULC)
- NATO Standardization Council (NATO/SC).

---

[18] CFAO 9-70 Special Commissioning Plan.
[19] CFAO 9-13 University Training Plan—Non Commissioned Members.

These organizations have, for a long time, provided leadership in Canada to help reduce the adverse effects of fire on the citizenry at large. Traditionally, membership includes persons representing a broad spectrum of interests in fire safety and fire protection with specific roles to play in research, codes and standards-writing, testing and certification of products, and development of fire protection equipment and materials.

Many members of the Fire Service have distinguished themselves by serving in key executive and administrative positions in these organizations. Those who served in this manner did so with the aim of making the Fire Service more effective and Canada a safer place in which to live.

The Council of Canadian Fire Marshals and Fire Commissioners

The CCFM&FC was formerly called the Association of Canadian Fire Marshals and Fire Commissioners (ACFM&FC). Directors of the CCFM&FC are the fire marshal or fire commissioner, as appropriate, of each province and territory, and appropriate representatives from the Government of Canada including the CFFM. Between meetings of the board of directors, the affairs of the council are managed by an executive committee headed by the president.

The mission of the CCFM&FC, as reflected in the letters patent, is to support its members in their efforts to minimize losses from fire by:

- advising on and promoting legislation, policies, and procedures pertinent to fire protection;
- participating in the development of codes and standards relating to fire safety;
- promoting fire-safety awareness;
- supporting the professional development of the Canadian Fire Service;
- arranging for the compilation and dissemination of national fire-loss statistics;
- identifying trends relative to the causes and the severity of fire;
- providing advice to accredited agencies involved in the certification and testing of fire-protection equipment, materials, and services relating to fire safety; and
- providing a forum for the exchange of information on fire safety matters.

Executive Positions

A number of Fire Service officers have served in the various executive positions of the ACFM&FC/CCFM&FC, including that of president (Lieutenant Colonel Lorne MacLean, 1984–85, and Lieutenant Colonel Hal Singleton, 1992).

The Canadian Association of Fire Chiefs—Mission

The CAFC is the national public service association dedicated to reducing the loss of life and property from fire, and advancing the science and technology of the fire and emergency service in Canada. The mission of the CAFC is to "strive for excellence in representation, information, education and service delivery to its members."

The Canadian Association of Fire Chiefs—Profile

CAFC is an independent, non-profit organization with a voluntary membership that was founded in 1908. In 1915 the association named itself the Dominion Association of Fire Chiefs. The name was changed in 1954 to the Canadian Association of Fire Chiefs. CAFC was incorporated in 1965 under the *Canadian Corporations Act* with its head office in Ottawa. An elected president leads the association and serves as chairman of its 23-member board of directors from across Canada. The president has the general charge of the affairs of the association. Fire Service fire chiefs have been members of the association for more than 30 years.

CAFC's activities are numerous and varied. For example, members participate in committees and represent CAFC with national and international organizations such as the National Research Council, the Transportation of Dangerous Goods Advisory Council, the Canadian Standards Association, the ULC, the NFPA, and the International Association of Fire Chiefs. The CAFC frequently performs an advocacy role to support and advance the interest of its members, the fire chiefs of Canada and, through them, the citizenry at large.

Involvement—Fire Service Members
Beginning in the early 1970s, members of the Fire Service demonstrated leadership by filling important seats within the CAFC and played leading roles in this national organization.

Office of the President
The following Fire Service officers have served as president of the CAFC:

- Major Bill MacDonald, on the staff of the CFFM, was elected as the 67th President of CAFC in 1975, in Mississauga, Ontario. This marked the first time in the history of the CAFC that a military FSO was elected president of this influential, national organization. He was re-elected in 1976 during the Vancouver annual meeting of the members.

- In 1984, Lieutenant Colonel Lorne MacLean, CFFM, was elected the 76th CAFC president in Vancouver, British Colombia.

- In 1989, Major Charles McNeil, Commandant of the Fire Academy, was elected the 81st CAFC president, in Montreal, Quebec.

- In 1992, Major Gaétan Perron was elected the 84th CAFC president in Calgary, Alberta.

Further to the above, in 1989, Major Marcel Ethier, retired FSO, was appointed executive director of the CAFC and secretary-treasurer of FPC. He remained in this position for ten years.

Fire Prevention Canada
The mission of FPC is to increase visibility and awareness of fire prevention and safety through education to the public directly and through individual fire departments. It is the primary Canadian source/supplier of fire-prevention materials and promotional awareness programs respecting fire safety. FPC has been managed by a board of directors comprising representatives from the CAFC and the CCFM&FC, which were its parent organizations. There are also directors from the private sector. Over the years, several members of the Fire Service have served on this board.

FPC is privileged to have Adrienne Clarkson, Governor General of Canada, as its patron. It is totally dependent on donations, grants and the sale of its fire prevention educational material to fund all of its numerous programs. To that end, FPC was incorporated in 1976 as a registered charity.

Other National and International Organizations
Several FSOs have served as president of the NATO Group for Standardization of Firefighting Equipment and Materials. For many years, meetings were held exclusively in Brussels, Belgium. In more recent times, meetings have been held in other locations. There has also been extensive involvement with codes and standards-writing organizations including ULC, NFPA, and CSA.

Fire Services Exemplary Services Medal[20]

Shortly after being elected president of the CAFC, it was decided to pursue the possibility of gaining formal recognition for the Fire Services of Canada similar to that which been extended to the Police Services. In this regard, the Chancellery of Canadian Orders and Decorations agreed to take an open-minded approach to the matter. After some preliminary work was completed, a committee of four that included the president and immediate past-president of the ACFM&FC and of the CAFC was struck. I was appointed its chair.

The committee began work on the development of the rules and regulations, under which the medal would be governed, and on the design of the medal. Proceeding in close co-operation with the Chancellery, the Fire Services Exemplary Service Medal, which recognizes the dedicated service of firefighters in Canada, was conceived. The birthing was another matter.

By this time, 1984 had passed into history and we were well into 1985. The critical next step was to gain the approval of the prime minister. To that end, a letter, with an accompanying package respecting the medal, was put together and forwarded. Considerable time passed without any indication as to how the request was received. Patience became the watchword. Care was exercised to guard against the potential of a negative reaction to any display of pushiness.

In midsummer, events began to unfold that eventually was to shoulder caution aside. The annual general meeting (AGM) of the CAFC, which in 1985 was being hosted by the City of Ottawa Fire Department in the Westin Hotel Congress Centre, was about to take place. Clearly it constituted an ideal time to break the news.

Perhaps somewhat selfishly, it was thought time to take a new tack. Accordingly, at the close of the first day of the AGM, Major Jack Henderson and I sat down to explore our options. It was decided to accept some risk by actively seeking a decision from the prime minister. Jack then took up the torch.

After brainstorming the options available, it was decided that he would call his reliable and trustworthy member of Parliament for Central Nova, the Honourable Elmer MacKay, who just happened to be a Cabinet minister and a personal friend of the prime minister. (He gave up his Central Nova seat to the prime minister during a by-election.) With luck on our side, and on the very first phone call, Jack was able to reach his MP, and after getting caught up on the current events from Central Nova, a request was made to have him contact the Prime Minister's Office (PMO) to determine the status of the file.

Within several hours, a phone call was received from the PMO stating that we must be mistaken as the file could not be found. With some persistence, Jack eventually convinced the PMO that the file had indeed been submitted to their office. Shortly thereafter Jack received a call from the PMO informing him that the file was found but had not been signed off by the prime minister. To complicate matters, the prime minister was at that very moment on his way to CFB Ottawa for the air force to fly him to Winnipeg for a conference and then on to meetings overseas. Not to be deterred, Jack made a second request of his MP to have him contact the prime minister and have him agree to sign off the file if it could be delivered to his aircraft before it departed.

cont'd on page 84

---

[20] Memoirs of Lieutenant Colonel Lorne MacLean (1951–87) and Major Jack Henderson (1961–92).

cont'd from page 83

Within a matter of minutes, the prime minister was contacted and agreed to approve the request if the file could be delivered to his aircraft prior to departure. With the help of the Ottawa Police Services and the Flight Operations Centre, the file was delivered to the aircraft (with only a slight delay in departure time) and the file was approved by the prime minister while en route to Winnipeg.

When confirmation was received that the request had been approved, it was hand-delivered to me. As luck would have it, I was at the podium chairing a plenary session of the members and was able to interrupt the proceedings to make the announcement. It was a pleasurable moment.

My role in getting the medal in place was generously recognized in December 1997, when I was presented with the Fire Service Exemplary Service Medal by His Excellency, the Right Honourable Romeo Leblanc, at a private investiture at Government House.

Fire Service Exemplary Service medal.

HONOURS and AWARDS

Introduction
Over the years, firefighters have received numerous awards and honours from a wide range of sources. A listing follows:

- Air Command Commander's Commendation
- Canadian Firefighter of the Year
- British Empire Medal
- Carnegie Medal
- Certificate of Meritorious Conduct
- Chief of Defence Staff Commendation
- Fighter Group Commander's Merit Award
- George Medal
- Medal of Bravery
- Order of Military Merit
- Queen's Commendation for Brave Conduct

Air Command Commander's Commendation

Corporal Ed Neufeld, October 1975
An incident that took place in 1974 underscored the versatility of the firefighter. It happened one night at CFB Cold Lake when a report was received that two aircraft were missing somewhere over the Primrose Lake range. Master Corporal Clackson and Corporal Ed Neufeld were detailed to accompany the crew of the

rescue helicopter in their search for the aircraft. For his courageous actions in rescuing the two downed pilots, Neufeld received the Air Command Commander's Commendation. (See Chapter X for details.)

Corporal Binnie, July 11, 1983
In recognition of outstanding service by Corporal Binnie, for his unusual fortitude, courage and resolve on November 16, 1982, in responding to the fatal crash of a Hercules aircraft at CFB Edmonton. With the main fire under control, although parts of the aircraft were still burning and explosions were still possible, he single-handedly removed the casualties and passed them to co-workers outside.

British Empire Medal

Terms
This medal replaced the Medal of the Order of the British Empire (1917–22) and had a military and a civil division. The medal was awarded for meritorious service which warranted such a mark of royal appreciation.

Civil Division
The BEM was not awarded to members of, or persons eligible for appointment to, any of the five levels of the OBE.

Military Division
The medal was awarded to subordinate personnel only; NCOs, petty officers and men, who were eligible for the military division of the various levels of this order. After the Empire Gallantry Medal was superseded by the George Cross, the BEM continued to be awarded for gallantry (but of a degree less than that required to earn the George Medal).

The British Empire medal.

Flight Sergeant Richard Albert Englebert, BEM, January 1, 1944
When an aircraft crashed on the aerodrome, turned over and sprayed gasoline into the cabin and surrounding area, Flight Sergeant Engelbert, with complete disregard for the grave danger of fire and explosion, forced his way into the cockpit, and although soaked with gasoline and suffering from the effects of the fumes, succeeded in cutting the unconscious pilot loose and removing him from the wreckage. Through his prompt action, courage and resourcefulness, this NCO was responsible for saving the pilot's life.

Flight Sergeant Engelbert was previously commended by the Chief of the Air Staff for a similar action. A letter of commendation dated May 19, 1942, from Air Marshal Breadner to Englebert, states that as a member of the fire crew at No. 6 Bombing and Gunnery School, Flight Lieutenant Englebert was instrumental in assisting in the extinguishment of a fire on Bolingbrook IV No. 9141, thus saving the aircraft from complete destruction. He was further praised for courage and resourcefulness. The incident occurred at Mountain View on April 16, 1942.

Flight Sergeant Englebert died of natural causes, March 25, 1960, while serving as a warrant officer in the RCAF.

Leading Aircraftsman Melvin Muir McKenzie, BEM, October 27, 1944

On a night in June 1944, an aircraft, while attempting to land, crashed into another, which was parked in the dispersal area and fully loaded with bombs. The former aircraft had broken into three parts and was burning furiously. Air Commodore Ross was at the airfield to attend the return of aircraft from operations and the interrogation of aircrews. Flight Sergeant St. Germain, a bomb aimer, had just returned from an operational sortie, and Corporal M. Marquet was in charge of the night ground crew, while Leading Aircraftmen McKenzie and Wolfe were members of the crew of the crash tender.

Ross, with the assistance of Marquet, extricated the pilot who had sustained severe injuries. At that moment, ten 500-lb / 27-kg bombs in the second aircraft, about 80 yd / 73 m away, exploded and hurled to the ground. When the hail of debris had subsided, cries were heard from the rear turret of the crashed aircraft. Despite further explosions from bombs and petrol tanks which might have occurred, Ross and Marquet returned to the blazing wreckage and tried in vain to swing the turret to release the rear gunner. Although the port tail plane was blazing furiously, Air Commodore Ross hacked at the perspex with an axe and then handed the axe through the turret to the rear gunner who enlarged the aperture. Taking the axe again, Ross, assisted now by St. Germain as well as by Marquet, finally broke the perspex steel frame supports and extricated the rear gunner.

Another 500-lb / 227-kg bomb exploded, which threw the three rescuers to the ground. St. Germain quickly rose and threw himself upon a victim to shield him from flying debris. Air Commodore Ross's arm was practically severed between the wrist and elbow by the second explosion. He calmly walked to the ambulance and an emergency amputation was performed on arrival at station sick quarters.

Meanwhile, Marquet had inspected the surroundings and, seeing petrol running down towards two nearby aircraft, directed their removal from the vicinity by tractor. Leading Aircraftmen McKenzie and Wolfe rendered valuable assistance in trying to bring the fire under control and also helped to extricate the trapped rear gunner, both being seriously injured by flying debris.

Awarded at the same time was the George Cross to A.D. Ross, the George Medal to J.R.M. St. Germain and M. Marquet, and the BEM to R.R. Wolfe.

Flight Sergeant René Noel Joseph Dupuis, BEM, November 14, 1944

Flight Sergeant Dupuis was in charge of the crew manning an RCAF marine craft, which carried out rescue operations after an RCN motor launch exploded at a naval jetty on September 7, 1944. The rescue of naval personnel from the water was carried out in close proximity to the burning craft, which carried full tanks of high-octane gasoline and battle complement of depth charges.

During these operations, smaller armament was being continually exploded by the fire, causing great hazard to Dupuis and his men. When a severely burned rating was pulled from the water into the rescue boat, Dupuis displayed great presence of mind and efficiency in carrying out adequate first aid until the arrival of the medical officer. Throughout the entire operation, he provided courageous and efficient leadership in the face of extremely difficult and hazardous circumstances. The outstanding ability and presence of mind displayed by this NCO was most praiseworthy.

Flight Sergeant George Constable, BEM, January 1, 1945

Flight Sergeant Constable received the BEM for his service during the war. He was born in Fairford, England, in 1901, and after immigrating to Canada, he became a City of Edmonton firefighter. Constable enlisted in the RCAF in 1940 and served in Toronto, Macleod, Regina and North Battleford. His wife learned of the award when her husband called Sunday night to wish her a Happy New Year. He said he could not think of why he should be getting an award, Mrs. Constable told the *Edmonton Journal* on the following Monday.

After a long and productive life, Flight Sergeant George Constable passed away on March 17, 2004, at the age of 102 years and 7 months.

Flight Sergeant Robert Benjamin Campton, BEM, June 9, 1945
Flight Sergeant Campton was on secondment to the US Army Air Force to assist in training personnel in firefighting. The course of instruction included assembling material built to resemble a crashed aircraft, which could then be saturated with hundreds of gallons of gasoline and ignited to provide firefighters with realistic training. On February 20, 1945, during one of these exercises, three American enlisted personnel were operating a fog nozzle in the midst of intense flame when the nozzle became jammed. Campton, realizing the danger, immediately rushed into the flames and succeeded in clearing the nozzle, obtaining a fog steam and cooling the fire. By his quick thinking and courageous action, at complete disregard to his own safety, he saved these personnel from becoming badly burned. In rescuing these men he received serious burns on his head and face.

Flight Sergeant William George Alexander, BEM, January 1, 1946
Flight Sergeant Alexander responded to the crash of a US Navy aircraft, and on arriving, found a fire burning above the gasoline-tank filler hole. Had the fire been driven down into the tank, an explosion with disastrous consequences would undoubtedly have occurred. With full knowledge of this existing condition, and without protective clothing or additional firefighting equipment, he approached the wreckage with a fire extinguisher, and with disregard for his safety, he put out the fire, thereby speeding rescue operations of the crew and minimizing danger to everyone at the scene of the crash. As station fire chief, Alexander at all times displayed outstanding ability and devotion to duty.

Flight Sergeant Robert Ian McKenzie, BEM, January 1, 1946
By January 1946, Flight Sergeant McKenzie had been at RCAF Station Uplands (Ottawa) for almost a year and had devoted time and energy for the improvement of conditions and efficiency of the fire department. McKenzie was an excellent NCO who continually and efficiently discharged his many responsibilities. His untiring efforts beyond the call of duty and outstanding ability were an example to all.

Flight Sergeant Allan Harvey Stotts, BEM, January 1, 1946
Flight Sergeant Stotts displayed exceptional skill and ingenuity as the NCO officer in charge of the fire department. His outstanding organizing ability in fire prevention coupled with a strong devotion to duty resulted in no fires at Station Torbay throughout his career. He was exceptional in his trade, above average in all respects, and a constant good leader of his men.

Flight Sergeant John Edward Blair, BEM, June 13, 1946
Flight Sergeant Blair showed himself to be especially outstanding in the performance of his duties. The supreme knowledge he had of his trade coupled with his vigour and enthusiasm in fire-prevention duties was an outstanding example to all ranks. The efficiency of his section contributed in no small measure to the absence of any serious damage by fire No. 1 Technical Training School over a period of over four years during the 1940s.

Flight Sergeant Arthur Paisley Copp, BEM, June 13, 1946

Flight Sergeant Copp performed his duties as fire chief in a most commendable manner, and was indefatigable in his efforts to prevent fire and fire hazards. His work was particularly outstanding because the danger of fire from personnel carelessness was greatly enhanced by the nature of RCAF Station Lachine (Montreal), the function of which was to serve as the embarkation point for personnel on overseas posting and later to receive repatriated personnel on their return. Copp used care and showed personal effort beyond the ordinary and had a fine record as a fire chief during the Second World War.

Acting Leading Stoker (FF) Gordon Morrison, BEM, 1946
Lieutenant (SB) William Carson, BEM, 1946
Commissioned Technical Officer (SB) Harold Coxon, Commendation, 1946

The Naval Fire Service was involved in several significant incidents, the most serious occurred in July 1946, and involved an explosion at the Bedford Magazines. Due to their outstanding performance during this incident, Acting Leading Stoker (FF) RCNVR Gordon Morrison was awarded the BEM for outstanding bravery, as was Lieutenant (SB) RCNVR William Carson, Command and Base Fire Chief, Atlantic Command. Commissioned Technical Officer (SB) Harold Coxon, Assistant Naval Fire Chief Halifax, received a commendation for his actions.

Certificate of Meritorious Conduct

RCAF Fire Department, Langar, England, Certificate of Meritorious Conduct, January 16, 1956
On December 17, 1955, at RCAF No. 30 Air Materiel Base Langar, England, a major fire occurred on the civilian side of the airport. The fire was in an aircraft maintenance hangar that held many 45-gallon drums and other-sized containers of paint solvents, thinners and the like, as well as Shakelton bombers that substituted for maritime patrol aircraft. The RCAF Fire Department was the first-in department and remained the sole department involved for an extended period. Although there was extensive damage that included one of the Shakelton aircraft, other aircraft and miscellaneous contents, and the structure was saved. Each member of the crew that was involved in the firefighting operation received a signed copy of the certificate.

Chief of Defence Staff Commendation

Major Lorne MacLean, December 1975
Corporal Donald Armstrong, December 1975

The commendation read, "In recognition of the courage and professional ability displayed at the scene of a fuel fire at Courtenay, BC, on the 28th September 1974. Their total disregard for their personal safety while carrying out the extremely dangerous task of stopping the flow of fuel from the tanks threatened by intense heat, and their application of expert knowledge in fighting the fire, averted a very near disaster." (See Chapter X for details.)

Fighter Group Commander's Merit Award

Sergeant R. J. Emrick, 1984

The award was presented for outstanding performance during OP EVAL 84 of CFS Beausejour, Manitoba, where Sergeant John Emrick was fire chief. It was the first time the award was presented to a Long Range Radar Station.

George Medal

Sergeant Douglas Stevenson, George Medal, July 10, 1956

On July 10, 1956, a fire occurred in a rail shipment of aviation fuel being unloaded at Montmedy, France.

The George medal.

Sergeant Stevenson, who was acting as deputy fire chief, performed two acts of bravery, which enabled the fire department to bring the fire under control and finally to extinguish it. The courage and unselfishness displayed by Stevenson, and the complete disregard for his own personal safety on this occasion, was in the highest tradition of the RCAF. In recognition of this, he was awarded the George Medal. (See Chapter X for details.)

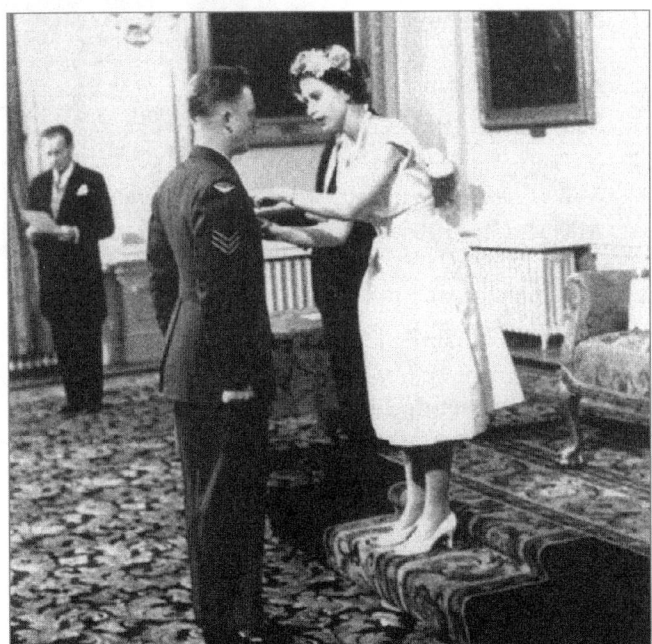

Queen Elizabeth II presenting Sergeant Douglas Stevenson with the George medal, July 10, 1956.

Medal of Bravery
FR1 David Crocker, MB, September 17, 1998
The Governor General presented FR1 David Crocker with the Medal of Bravery for the daring rescue of Mrs. Debra Wells and her children from her family's third floor apartment after fire erupted in the living room in Dartmouth, Nova Scotia, on January 20, 1997.

Crocker also received the Carnegie Medal which is awarded to one who, at the risk of his or her own life saves or attempts to save the life of another person. He was also selected as the Canadian Firefighter of the Year by the CAFC. (See Chapter X for details.)

Medal of Bravery.

FR 1 Mark MacKean, MB, February, 2005
FR 1 Michael Payne, MB, February, 2005
On December 30, 2002, a team of 10 people rescued a teenager who was trapped under a collapsed wall during renovations being made to a century-old home, in Stayner, Ontario. Although aware of the risk of the two-storey structure completely collapsing, the crew entered the building and, using minimal temporary bracing, secured the remaining walls. Through rocks and bricks obstructing their way, they managed to reach the young victim, pinned under a pile of rubble at the bottom of the basement stairs. Using their bare hands and makeshift tools, the team laboured for 40 minutes in the dusty and unsafe environment, removing the concrete piece by piece, to free the severely injured teenager from his precarious position and to bring him to safety.

Order of Military Merit

Its Historical Roots

Since ancient times, it has been the practice for those in the military who distinguish themselves by their exemplary courage or merit to be awarded some mark of distinction. Through the centuries, these marks of honour have taken many forms, from the laurel wreathes with which the early Greeks and Romans crowned their heroes to special presentation swords and other weapons. In more recent times, it has become customary to bestow on those to be honoured, a special badge in the form of a medal, star or cross, suspended from a ribbon and worn around the neck or pinned to a coat, to distinguish them from their comrades.

Its Canadian Roots and Application

During Canada's Centennial Year, in 1967, the Order of Canada was established as a means of honouring those who rendered meritorious services to the nation. It was also decided to include in Canada's Honour System a means of recognizing conspicuous merit and exceptional service by the men and women of the CF, both Regular and Reserve. Accordingly, on the first day of July 1972, the Order of Military Merit (OMM) was established. The OMM limits appointments to one-tenth of one percent of the average number of persons who were members of the CF during the preceding year.

The OMM confers no special privileges on its members and brings them no monetary rewards. It is a society of honour with three degrees of membership: commander, officer and member. The members of the Fire Service who have been appointed to the OMM are as follows:

- WO R.S. Anderson, MMM, CD, 1973
- MWO Ron Rockey, MMM, CD, 1976
- MWO Vern Robison, MMM, CD, 1978
- CWO Bob Hotston, MMM, CD, 1980
- LCol Lorne MacLean, OMM, CD, 1985
- CWO Jim Munro, MMM, CD, 1989
- CPO 1 John Daley, MMM, CD, 1992
- CWO Bob Morrison, MMM, CD, 1993
- MWO Don Armstrong, MMM, CD, 1995
- MWO Doug Goodings, MMM, CD, 2004

Order of Military Merit medal.

Queen's Commendation for Brave Conduct

Sergeant Leandre "Pat" Finnigan, Queen's Commendation for Brave Conduct, June 29, 1968
At 0300 hours on December 6, 1967, the fire department at CFB Ottawa (then RCAF Station Uplands) received a fire alarm from the Motor Transport Garage. On arrival at the scene, Crew Chief Sergeant L.J. Finnigan, who was in command, observed that an explosion had occurred and that a fire

was burning below floor level in the equipment room. With creditable dispatch, he assumed control and proceeded to direct his crew in fighting the fire and directed other personnel in removing to safety the vehicles parked in the garage.

During this entire incident, but most apparently during the latter phases, Finnigan displayed attention to duty and high regard for the safety of his personnel while knowingly risking further injury to himself with possible grave consequence. (See Chapter X for details.)

FATALITIES—LINE of DUTY

Introduction
This record pays special tribute to the firefighters who lost their life in the line of duty. A great deal of research went into developing this honour roll of those who lost their lives in the service of others. Even so, it is recognized that there may be oversights. Should this be the case, it is both unintentional and deeply regretted.

The Honour Roll
- Flight Sergeant Milner, RCAF Station Whitehorse, Yukon, 1947
- Leading Aircraftsman Fisher, RCAF Station Fort Nelson, British Columbia, late 1940s
- Leading Aircraftsman Hawley, RCAF Station Gimli, Manitoba, 1950
- Leading Aircraftsman Bolong, RCAF Station Goose Bay, Labrador, early 1950s
- Private Montreuil, Fire School, Borden, Ontario, February 3, 1976
- Firefighter 1 Huges, CFB Goose Bay, Labrador, 1990

BURNISHING the TIES that BIND

Overview
Over the years, firefighters have felt the need to maintain and strengthen the enviable cohesiveness that is a hallmark of the Fire Service and have developed various means to that end. Some of these vehicles for social interaction are described on these pages.

Box 1701 Club[21]
The story of Box 1701 Club is told here as it was described by Major Phil Brown, with minor editorial adjustments to facilitate its presentation. Although the exact date that the club was formed was not firmly established, it can be stated with some confidence that it was in the mid to late 1950s. This is a good point to pick up Phil's narrative.

> At that time we, of the Air Force Fire Service, were vaguely aware that government departments also had fire-protection requirements, but outside of our opposites at the Navy and Army Fire Marshal's offices, knew little of them. The various levels of service messes had proved that social intercourse produced contacts that often led to the sharing of knowledge and the solving of problems in our primary field of endeavour.
>
> In recognition of the potential for good that this concept embodied, I proposed a monthly social gathering at an officers' mess of all fire-protection personnel in the Ottawa area that could be contacted. The initial reaction from my colleagues on the job was negative. Despite this, the first meeting was called. It was attended by the staffs of the

---

[21] Major Phil Brown (1947–75), *The Royal Canadian Air Force Fire Service 1939–1975, A Subjective History* (unpublished).

fire marshal's offices of the three military services. Discussions within this group led to the development of a list of other fire-protection officers in the Ottawa area. This was reflected in the attendance at the next meeting when personnel from the National Research Council, the Department of Transport and the office of the Dominion Fire Commissioner came forth. From this point onwards it snowballed until we could count on 40 or 50 to be in attendance each month.

Eventually the club encompassed several government departments, including the RCMP. There was no formality. Members bought their drinks at the bar and sat in a reserved section to socialize amongst themselves. The contacts that were established served well, repeatedly proving their worth.

The club got its name from a chance remark of my good wife in response to me mentioning one Thursday that the next day was the monthly get together of the Club and expressed the wish to tag a suitable name on it. She knew that we quit work at 1700 hours and said, "Why don't you call it Box 1701." I replied that 1701 was a ridiculous number for a fire-alarm box, since a ticker tape alarm system could not produce a zero, and her reply is forever engraved in my mind: "It's a ridiculous club anyway." The next day the members endorsed the name; it stuck for many years. A wooden alarm box was made and a sign-in book placed inside with the following dedication:

"To the Box 1701 Club: a convivial gathering of government fire protection officers with the avowed aim of better inter-departmental relations through social contact, this register and symbol is respectfully presented. May the record maintained therein reflect the true spirit of camaraderie present in our chosen field."

Curling Bonspiel
The Firefighter Bonspiel started as a fun day for all Air Force firefighters at our European NATO bases. The first bonspiel, which took place in February 1965 at No. 2 Wing, Grostenquin, France, was organized by Flight Lieutenant John Cowell, assisted by "Jeep" Collins[22] both from ADHQ Metz, France, where John was fire marshal, and Wally Wishart from No. 3 Wing, Zweibrücken, West Germany. Curlers from ADHQ, Metz, No. 1 Wing, Marville, France, No. 3 Wing, Zweibrücken, West Germany, and No. 4 Wing, Baden Söllingen, West Germany, joined those already in place at No. 2 Wing. The day was so successful that it was decided that if at all possible it should be an annual event.

In February 1966 the second bonspiel was also held at No. 2 Wing. This was even better attended and set the stage for expansion of the event. There was no bonspiel in 1967 because of event timing and ice availability.

The third bonspiel was held in January 1968 at RCAF Station Uplands (Ottawa), the first held in Canada. By this time, John was serving as fire chief at RCAF Station Uplands where he, with support from other members of the fire department, undertook to co-ordinate the many tasks involved in getting the word out to firefighters across Canada and Europe, making arrangements for accommodations for visiting teams, booking ice time, and so on. Invitations were only sent

---

[22] Neither the rank or position of "Jeep" Collins and Wally Wishart at the time of the first Firefighter Bonspiel or upon retirement could be established.

to those RCAF stations within easy road travel of Ottawa. This bonspiel was thought to be the real test as to its sustainability as an annual event. Again the response was most favourable, and it was decided that for the fourth bonspiel, invitations would be sent to all bases where military firefighters were on strength.

The growth of the event continued and for the fifth bonspiel, held in February 1970, the invitation list was expanded to include all military and civilian firefighters. It was, in retrospect, a big step forward, as teams came from across Canada and Europe. Records show that at this point a rather large transition took place, as the number of rinks competing reached 28 in 1973. Over the next 30 years, there has been a gradual reduction in the number of rinks attending coincident with the number of firefighters and the availability of military air transport.

The bonspiel has generally been hosted by Fire Service personnel at Borden, Trenton or Uplands. However, over the last few years, CFB Trenton has evolved as the official host of the event and does so with great success each year. The latest being the 38th Annual Bonspiel in November 2003. More of the Firefighter Bonspiel's history can be found in the Curling History Albums created by the founder. These are held in trust by the CFB Trenton Fire Department and are displayed for all to enjoy at each bonspiel.

In every case, those involved gave much of themselves to ensure it was always an event that generated pleasant memories. After all the rocks are curled, the team that emerges triumphant is presented with the "Foam Bowl" by its sponsor Flying Officer (Retired) Doug Stevenson, president of Douglas Fire Safety Systems. This they may hold until the following year, when it must be exposed to the skilful efforts of their competitors.

From its inception, the event was an unqualified success. This should not be surprising since it was and remains a vehicle for social interaction that could only

F/L John Cowell in discussions with Sgt Rene Dupuis.

The 1969 Curling Bonspiel Champions. Doug Stevenson presents the Foam Bowl to Skip F/L John Cowell, WO Jim Lockhart, Cpl Ernie Strocel and Cpl Bill Rowat.

exist in an atmosphere of relaxed jovial competition, and exist it has. The Firefighter Bonspiel has the longest uninterrupted run of any military curling bonspiel that is held annually. With the passing of the founder, Flight Lieutenant John Cowell, on July 27, 1999, the event was officially renamed the John Cowell Memorial Bonspiel.

Golf Tournament[23]

The Firefighter Golf Tournament originated in the early 1970s and has subsequently been held on an annual basis. Based on what can be determined from very sketchy records, either 1973 or 1974 is thought to be the most accurate founding date. In the beginning, Major Art Haggart, who was the officer commanding the Fire School at the time, was of the view that, because there was an annual firefighter curling bonspiel that brought firefighters together and was widely enjoyed during the intemperate periods of the year, there should also be a summertime event that could similarly capture and enhance the already demonstrated camaraderie that dwells within the Fire Service. After a period of brainstorming with other members of the Fire School staff, a golf tournament was selected to fill the need, and the structure of the Annual Firefighter Golf Tournament was developed.

The annual firefighter golf tournament is open to all firefighters, whether serving or retired, and to fire service, fire support personnel, such as MSA, Levitt Safety and Acklands Grainger, and others. The event is always held in Borden with the Base Fire Department and the CFFA staff alternating as hosts on a year over year basis. The event is also usually held the last Saturday in August; however, on some occasions, due to golf course bookings and the like, the event has been held in June.

Western Canada Reunions

The first Western Canada Reunion was held in Red Deer, Alberta, in the summer of 1994. The next reunion was held in Edmonton, Alberta, in 1996, entitled Smoke Eaters '96, and was organized by Master Warrant Officer (Retired) Don Teed. Subsequently, the event has been held at various locations at two-year intervals. Each event is given an individual title. The format includes a meet-and-greet on Friday evening, a mix of activities on Saturday, with a social gathering and dinner in the evening. The final activity is a godspeed barbecue in the early afternoon, Sunday.

Eastern Canada Reunions

Fire Service members in the Annapolis Valley of Nova Scotia have been holding reunions for over 30 years. The first of the reunions was started in the earlier 1970s by the late Major Vern Collins (1945–76), then fire chief of CFB Greenwood. He started by inviting the single personnel to his house for a Christmas gathering, and later expanded it to include the local retired members and their spouses. This annual event was sometimes held in local eating places in the Annapolis Valley area, but it was always home-based at Vern's place until his wife Bea passed away on October 28, 2002.

To accommodate everyone, a summertime event was included, which from 1974 through 2002 was always held at Sergeant "Bud" Yeomans' (1944–70) cottage at Lake Ramsey, north of Bridgetown, Nova Scotia. These events would attract anywhere from 15 to 45 people. At some point it was decided to plan a mess dinner event, which would include all members in the Atlantic area. The first one was hosted by Fire Chief Paul Olster at CFB Cornwallis, in 1992, and the second in 1993. The third and fourth were held at CFB Greenwood in 1995 and 1996. These mess dinner reunions generally had

---

23 Memoirs of Warrant Officer Glen Herman (1954–93), Warrant Officer/FR5 Bob Labbe (1954–91), Lieutenant Colonel Bob Maxwell (1956–93), Master Warrant Officer/FR 7 Ian Morrison (1951–95), Master Warrant Officer/FR4 Ron Rocky (1943–88), Master Warrant Officer Don Teed (1951–80) and Chief Warrant Officer Joe Walker (1962–90, 1998–99), as compiled by Captain Gary Oliver (1979+).

from 65 to 193 in attendance. Even when the mess dinner reunions were held, the two annual local events in July and December have always taken place. Although over the years the numbers attending have dropped some, due in a large part to the closure of CFB Cornwallis, the events are still being held.

The Round-Up

By way of background, for a number of years a core group of serving and retired firefighters in the Ottawa area had habitually met for lunch on Fridays to socialize and perhaps do a bit of work. On one such day, the idea of casting a wider net to include others within easy commute of Ottawa was bandied about. Nothing concrete was done at the time, but after a rather extended gestation period, it was decided to give it a try.

Accordingly, the process of getting names, phone numbers and mailing addresses got underway, and the first such get-together was scheduled for the fall of 1997. It being in the fall, the practice of ranchers gathering up "strays" each autumn came to mind and it was dubbed the Fall Round-Up. As time moved along, it is simply referred to as the Round-Up.

In addition to providing an opportunity to socialize and enjoy a buffet luncheon, it provides an opportunity to present an honour roll of all firefighters who have answered their last alarm during the preceding year. Since its inception, it has been held annually, and attendance has increased, year over year.

The Fire Service "Online"

With the introduction of the computer and the Internet into most Canadian homes in the 1990s, it did not take long for some members of our Fire Service to establish a Web site for their profession. In February 1999, two retired firefighters, Master Warrant Officer Paul Landry and Chief Warrant Officer Joe Walker, met by chance

Some of the participants in the 2003 Round-Up.

F/L Keith Potter and F/L Archie Graham in the foreground, Maj Marcel Ethier, Capt Hamilton and WO John Colbourne in the background.

Maj Andy Beaudin, CWO Steve Shand, and CWO Yvon Serre.

MWO Frank O'Meara, LCol Lorne MacLean, and WO Tom Ronan.

on the Internet, and out of a conversation between them, the FireHouse651.com concept was born.

The Web site was launched on April 5, 1999, with an initial mission to locate all the Canadian military fire service members who were online around the world. Captain Al Rau and Master Warrant Officer (Retired) Don Teed joined the team as directors, and working together, they maintain over 800 Web pages dedicated to the profession and their fellow military firefighters. To date, they have registered contact information for over 1,300 members in the databank, and the membership grows each month. They also record the names of deceased firefighters in a "We Remember Them" memorial garden.

The Web site is privately owned and funded through donations from a few in the membership and several corporate supporters. The main features are discussion forums, member's databank, firefighter memorial, photo library, guest book, e-mail accounts, the newsmagazine *The FireFlyer*, and many other features of interest to the Fire Service. They have also established and maintained a Web site for the Corps of Canadian Firefighters that served in England during World War II.

FireHouse651.com aims to provide a common place on the Internet for serving and retired military and civilian firefighters of Canada's Armed Forces to register their contact information. The aim is to provide a service that will enable the members to reunite with their old friends and for them to establish new friendships.

The services also include:
- collecting and publishing the historical material of the Fire Service;
- providing and maintenance of a Memorial Cemetery within FireHouse651.com;
- providing information on Fire Service matters and topics of general interest;
- providing Internet e-mail services and other features to the members; and
- providing support to organized Fire Service functions and reunions.

The managers of the FireHouse651.com Web site: MWO Paul Landry, CWO Joe Walker, MWO Don Teed, and Captain Al Rau.

CHAPTER VI

PROFESSIONAL DEVELOPMENT

The First Schools of Firefighting
In 1941 and 1942, the Army had firefighting training schools at Camp Borden, Ontario, and Camp Chilliwack, British Columbia, that conducted brief courses on firefighting. Concurrent with this, the Air Force began operating its first organized firefighting school at RCAF Station Mountain View, Ontario, a satellite airfield approximately 10 miles / 16 km from RCAF Station Trenton. Flying Officer W. McCallum, a recruit from the Toronto Fire Department, became its first chief instructor. With only a small nucleus of instructors, the task of training newly inducted recruits commenced in May 1941, with the firefighting exercises taking place at RCAF Station Trenton. The course size varied, but it was not uncommon for a course to have 50 or 60 students.[1]

In 1944, the first Navy fire school was established in Esquimalt, British Columbia, under the direction of Chief Petty Officer Gordon Lay (FF).

The training program carried out at RCAF Station Mountain View and RCAF Station Trenton was a resounding success. The fledgling school was quick to meet the immediate needs of the Air Force and graduated a considerable number of firefighters in a relatively short time. Unfortunately, the school became the victim of its own success when senior Air Force officers considered the problem of firefighter manpower shortages solved and closed the facility.

Critical Shortage of Apparatus
ARFF vehicles, airfield crash tenders as they were

---
[1] Flight Lieutenant John W. Colwell, CD, *A History of the RCAF Firefighter* (unpublished.).

called at the time, were in great demand, with most being transported straight from the manufacturer to the flying-training stations. As a consequence, instruction on this type of equipment had to wait until the newly graduated students arrived at their parent unit. In a well-meaning attempt to introduce students to the crash vehicles, they were marched over to the operational side of the airfield, where they got a long-range look at a crash vehicle, through a fence! Since the school did not have any crash trucks, this was as close as the students got to this highly important piece of equipment until they arrived at an operational unit.

Employment of Course Graduates
In July 1940, outside the scope of the BCATP, the British government requested that 14 RAF elementary flying training schools be moved to Canada, which the Canadian government accepted.

Many of the trainees were members of the RAF who, upon qualification, went on to serve at RAF stations in Canada. RCAF members were normally posted to stations in Canada, although 25 did go overseas, albeit in error.[2] British authorities had asked for firemen; however, they were looking for personnel trained to work with pressure boilers. AFHQ misinterpreted the request to mean firefighters and promptly sent 25 firemen to England. The mistake was realized upon their arrival, but once there they eventually joined stations supporting No. 6 Bomber Group. Later they served in France and Germany, providing crash protection on airfields and manning fire vehicles abandoned by the retreating enemy. The firefighters assigned to the European theatre found themselves in hazardous situations, as they were employed fighting fires in downed bombers, with all the associated risks and hazards from exploding ordnance and fuel.[3]

Addressing an Urgent Need for Firefighters
To address an urgent requirement, the fire school at RCAF Station Mountain View re-opened in late 1946. Flight Sergeant A. "Bull" MacFadyen was appointed chief instructor and was assisted by Flight Sergeant "Jock" Smith and Corporals Cowell and Wilson. The school again became part of the Composite Training School (KTS). To avoid confusion between Composite Training School and an aircrew Conversion Training Squadron, the letters KTS signified Composite Training School, when abbreviated.[4]

Courses were scheduled every three weeks. Two of the first courses were allotted to civilians with firefighting background, while the remaining courses were made up of recruits without previous fire service experience.

Training was three weeks duration with emphasis placed on developing an understanding of theory of fire behaviour and on use of portable fire extinguishers. Practice firefighting evolutions were carried out using the two in-house LaFrance pumpers. These trucks were built on an international chassis with an open cab, which was the custom of the time. Although this open cab design proved a little impractical for northern climates, it remained a popular vehicle. Equipped with a rotary-gear positive displacement pump, it was capable of producing 600 G/min / 2700 L/min of water at 120 psi / 830 kPa. Students practised fire operations, including drafting and pumping exercises, at the seaplane hangar on the Bay of Quinte.[5] Hangars and surrounding buildings became practice sites for ladder drills and accompanying hose evolutions.

The haste to fill the vacant positions meant that some courses became seriously overloaded, one course having 65 students. Exigencies of the service necessitated the speedy training program.

---

2   Ibid., p. 9.
3   Deveau, *The Confusion of War* (unpublished recollections, 1956).
4   F.J. Hatch, *Aerodrome of Democracy* (Ottawa: Canadian Government Publishing Centre, 1983), p. 118.
5   Major Phil Brown (1947–75), *The Royal Canadian Air Force Fire Service 1939–75, A Subjective History* (unpublished narrative, 1992), p. 3.

The Various Locations of the Fire Schools

The history of military fire service training goes back to World War II. At that time, the Air Force identified the need for a fire service and established a school at RCAF Station Mountain View in the early 1940s to address the need. In 1947, the school moved to accommodations at RCAF Station Trenton, where basic firefighting courses were conducted until late 1949. It is believed that the training facility at Trenton was closed for a brief period and then reactivated in 1950 under the leadership of Warrant Officer 1 Dave Lefebvre. Some time later, due to the large volumes of black smoke produced by firefighter training, the fire school was found to be incompatible with flying operations and was moved again. This time to RCAF Station Aylmer, Ontario, where it remained until 1956, when it moved to its permanent home in Camp Borden.

Establishing and Stabilizing at Borden

Under the leadership of its officer commanding, Flight Lieutenant A. "Bull" MacFadyen, the fire-service training school became firmly established, continuing today, as "Home of the Armed Forces Firefighters." In the beginning, three First World War-era aircraft hangars were quickly converted to classrooms, vehicle storage garages and equipment maintenance shops. Over the years, various mock-ups that could simulate a burning aircraft fuselage were obtained, a structural fire-training tower was built, and a separate building was constructed to house fire-alarm systems, sprinkler systems and other fixed-pipe fire-suppression systems for use as training aids. It would, however, be the better part of a quarter of a century before more advanced and sophisticated facilities, where form more accurately reflects functions of the CFFA, would be developed and come into use.

Change and Stability

In 1970, the Fire School's name was changed to FFTC after it was designated an integral part of the CFSAOE. A period of stability and consolidation within the Canadian Forces Training System that ensued remained in place until the 1980s when early moves were made towards unifying trades training. At that time there were studies underway to dismantle or split up CFSAOE. As well, NDHQ and the base commander, Brigadier General Jim Hansen, pushed the creation of a separate and distinct firefighting training facility at CFB Borden. With the break-up of the CFSAOE organization in 1985, the FFTC became the CFFA.

Updating and Broadening the Professional Development Program

In 1972, Master Warrant Officer Frank McCollum of CFSAOE Standards Company for FFTC assembled a writing team from the field to review and upgrade the entire firefighting training program. The team of Warrant Officer Don Bruton, Sergeant Joe Walker, Master Corporal Jack Kearley, Master Corporal Mike Savard and Corporal Marcel Ethier spent four months researching and developing new training methods and reference materials that were written into Training Systems Headquarters' new performance objective/ enabling objective method of teaching. The work of this group and their introduction of an updated Course Training Plan had a major impact on the training programs at FFTC and firefighter training in the field.

Tragedy Strikes

A vehicle accident occurred on February 3, 1976, in the FFTC training area that resulted in the death of Private Paul Montreuil, while attending a qualification level three (QL-3) basic firefighting course. The tragedy cast a pall over the FFTC and, in the broader sense, the Fire Service as a whole.

Training Evolves and Continues

Trades-training continued to be the major commitment for the CFFA in the mid 1980s and early 1990s. The recruiting rate decreased in the mid 1980s when unit establishments became filled, which allowed the trades-training backlog to be reduced. Broader reductions during the latter part of this period, which included the closure of bases and the concomitant loss of fire stations, reduced the total number of firefighters from approximately 700 military and 700 civilian to 450 and 475 respectively in each group.

For quite some time, due to the ongoing demand to conduct training, there had been little opportunity to develop a full set of career courses for firefighters. Consequently, one of the key tasks which the Academy faced after 1987 was to establish more specific training standards, syllabi and training manuals. A complete rewriting of the trades-training standards and performance objectives resulted in a progressive curriculum beginning with Apprentice, Qualification Level 3 (QL-3), Journeyman, (QL-5), Supervisor/Commander, (QL-6) and Manager/Commander (QL-7) courses keyed to the appropriate nationally and internationally recognized fire protection codes and standards, including those dealing with emergency operations respecting structural firefighting, casualty extrication and rescue, and aircraft rescue and firefighting, with the overarching aim of imparting a broad range of fire service general knowledge.

To complement this, specialty courses were developed on fire prevention and life safety inspector, and on hazardous materials at both the technician level and specialist/manager level. All programs were certified and accredited by the International Fire Service Accreditation Congress (IFSAC) of Oklahoma State University, which audits teaching institutions to ensure fire service training meets the challenging NFPA Standards allowing recognition between jurisdictions of qualifications attained at an accredited teaching entity.

The Fire Inspector Course was introduced into the Fire Service training in the mid 1970s. A writing team was assembled in Winnipeg in the winter 1973–74, with chairman Major Brian Crow and board members Warrant Officer Lloyd Howlett, Sergeant Joe Walker, Mr. Jerry Berube and Warrant Officer Bob Briggs. The team wrote two courses: a Basic Fire Inspectors for corporals and below and an Advanced Fire Inspectors course for sergeants and above, which eventually were combined into one package. This training program was a major step forward for the Fire Service members in gaining the knowledge required to become a fire inspector. The Fire Inspector Course was removed from the training programs when all the training quotas were met, but was re-introduced with accreditation credits in 1999. Other courses were updated for the IFSAC certification team.

In the late 1980s, a re-certification training program for firefighters with responsibilities for aircraft rescue and firefighting was established when military bases and units were required to close their live fire training areas because of environmental concerns. This training is keyed to the requirement for firefighters to have live fire training to maintain skills as recommended by NFPA[6] and required by Standard NATO Agreement.[7] The course was originally aimed at re-certifying the driver/operators of crash trucks. The course has evolved to one in which team concept is stressed, which permits the students to practice as a unit while drivers/operators are evaluated. (See details at Appendix F.)

Students to be Demonstrably Physically Fit
The subject of physical fitness programs has been dealt with in some detail in Chapter V—Personnel. Coverage here is limited to the application of these programs in the training environment.

---

[6] NFPA 405 Recommended Practice for the Recurring Proficiency Training of Aircraft Rescue and Fire-Fighting Services, 1999 Edition.
[7] Stanag No. 7145 Minimum Core Competency Levels and Proficiency of Skills for NATO Firefighters.

In the early 1960s, the RCAF firefighter students were required to meet the fitness requirements of the RCAF 5BX fitness program, which was not firefighter-specific. Although improvements were made over the years, firefighters were still not up to a fitness standard that would permit them to safely and effectively participate in firefighting and rescue operations, irrespective of whether or not these were actual emergencies or simulated emergencies in the training environment.

Clearly, it was a logical requirement for the Fire School / Fire Academy to make this determination before conducting strenuous exercises. In the late 1970s and early 1980s, a set of standards was developed, which incorporated items to measure: cardiovascular efficiency; the ability to push using various firefighting tools; the ability to pull on hose, ropes and other equipment; balance, in order to perform in circumstances where the footing might be unstable; and strength sufficient to carry a victim in emergency circumstances. Beginning in the late 1970s and through much of the 1980s, this was done by evaluating students against a firefighter-specific test standard. Some years later, a second firefighter-specific program was developed and testing against this standard began in 1999. The test requires firefighters in full gear to perform ten tasks that simulate general firefighting duties.

The Transition into an Academy

As a result of the split-up of CFSAOE, the CFFA was established on September 1, 1985, as a component of the Canadian Forces Training System (CFTS) Headquarters, located at CFB Trenton.[8]

The formation of an academy was a significant and almost seamless step forward for the Fire Service, as the organization went from a training company with an officer commanding (OC) to an independent academy with a commandant. The initial organizational structure and key personnel included: Commandant, Major Charles McNeil; Chief Instructor, Captain Dick Sadler; Academy Chief Warrant Officer, Chief Warrant Officer Joe Walker; and Chief Clerk, Warrant Officer Gord Sabourin.[9]

Flag-raising ceremony. In 1987, BGen Jim Hanson raised the flag of the CFFA for the first time.

Igni Obstare—Standing Against Fire.

---

[8] Authorized by Message DMCO 238 061800Z, September 19/85.
[9] History of the CFFA, Firehouse651.com.

In the late 1980s, Queen Elizabeth II approved and signed the Academy's official crest. The CFFA's motto, *Ingi Obstare*, which translates as Standing Against Fire, is displayed at the bottom of the crest. It was a very proud moment when, in 1987, a new Fire Academy flag was unfurled by Brigadier General Jim Hanson, Base Commander of CFB Borden.[10]

With the construction of facilities nearing completion, a great challenge for the Academy became a stand-alone academic institution offering a full range of courses for firefighters of all ranks, as well as providing instruction on the role of the Fire Service to military engineer commissioned officers, as part of their basic training. With these developments, the CFFA responsibilities exceeded those of the former school, conducting a complete spectrum of planning, organization, instruction and administration of the courses as a component of the CFTS.

Getting the Academy Up and Running

CFFA staffing evolved from the former Fire School, which was staffed by trained and experienced professional Fire Service personnel. When the change was completed, the establishment stood at 36, having been increased to cope with the expanded

CFFA's inaugural staff, 1985.
Back Row: Sgt Kelly, Sgt Bedard, MCpl Millman, MCpl Paquet, Cpl Larocque, MCpl Hebert, Cpl Goodwin, MCpl Hinds, Sgt Jobe, Cpl Oliver, Cpl Lamoureux, MCpl Chamberlain, MCpl Normand, Cpl Cyr and MCpl Rioux.
Centre Row: Sgt Landry, Sgt Goddard, Cpl Beloslaudzew, MCpl Vachon, Cpl Braze, MCpl Young, Sgt Nurse, MCpl Reid, Sgt Millar, MCpl Robertson, Cpl Bishop, Sgt Tremblay, MCpl Beauchesne, MCpl Wiken, MCpl May and Sgt Brouillard.
Front Row: WO Sabourin, Cpl Englehardt, Mr. Halfpenny, WO Simmons, WO Neufeld, Capt Sadler, Maj McNeil, Mrs. Hodgins, CWO Walker, WO Tiernan, Mr. Hamel and Cpl Dusault.

---

10 CFFA Annual Historical Report, 1987.

responsibilities associated with being a separate school. Firefighting instructor positions were identified and defined to meet the expanded training requirement.

The main task CFFA faced was to establish more specific standards for all new courses and to write the required syllabi and training manuals. This was accomplished by using a completely revamped Firefighting Trades Training Standards and by borrowing the experience from other organizations, such as NFPA, International Fire Service Training Association (IFSTA), Canadian Fire Investigation School (CFIS), and Canadian Forces School of Instructional Technique (CFSIT), and observing USAF firefighting training methods at Tyndall, USAF Base, in Florida.

Background—Academy Infrastructure

The CFFA, and to a greater extent its precursors, had for 40 years been located on the hangar line in Hangars 3, 4, and 5 at CFB Borden. Consequently, it was with great pride that CFFA finally had the official opening of the new Headquarters Building on September 13, 1996, and changed its location from Hangar 3 to Building A-256.[11]

Developing New Facilities

As funds became available during the 1980s and 1990s, the Academy expanded physically, with new buildings, modernization and upgrading of the training areas, and the acquisition of new equipment. The improvements included Canada's largest aircraft mock-up. The level

Staff of the CFFA, 1998.
Back Row: Cpl North, MCpl Gelinas, Sgt Belanger, MCpl Hebert, Sgt Pilot, Cpl Lesard, Sgt Emms, MCpl Brophy, Cpl Dunn, Sgt Ross, Cpl Leduc, Sgt Prochilo, MCpl Morris, Sgt Turgeon, MCpl Grant and Sgt Sawkins.
Front Row: Mrs. Chase, WO Laroche, WO Pennell, MWO Reid, CWO Serre, Maj Colledge, CWO Walker, MWO Lafond. WO Ladoucer and Mrs Crocker.

---

11   CFFA Annual Historical Report, 1997.

of interest in the mock-up was sufficient to generate a contest to give it a name. At the official opening ceremonies in 1986, it was dubbed the Mocking Bird,[12] as proposed by Private Joyce Halleran.

The Mocking Bird.

The completion of updated aircraft mock-ups, two additional concrete training towers, a modern fire simulator for training senior Fire Service members, and several new maintenance and training structures have gone a long way to satisfy the long-term requirements of the CFFA. These first-class facilities were the vision of far-seeing senior firefighters, who were determined to make CFFA the finest fire-training institution in Canada. It was a credit to their drive, foresight and pragmatism that these fine facilities were ultimately established.

Fire-training areas have always been a problem, due to smoke plumes from fires involving various types of fuel. Drainage of unburned fuels and fire-suppressing chemicals have been a frequent source of complaints from surrounding communities and of concern for the military. To satisfy the critical requirement to train students in the strategies and techniques of suppressing fire and conducting rescue operations in menacing conditions, a firefighting training area was developed to sharply reduce adverse impact on the environment. It opened in 1993.[13] The training area provided a hands-on realistic environment, which included a five-car train wreck, three full-size aircraft mock-ups, two building mock-ups, and a wide variety of training facilities spread out on a 25-acre site.

A Catalogue of Facilities

The details of the Academy's infrastructure, as catalogued in 2003, are listed below:

- a 25-acre / 10-ha training area with a drainage system designed to preclude environmental damage
- a building housing the CFFA HQ, administrative and instructor offices and classrooms
- an indoor simulator capable of producing complex structural firefighting and aircraft rescue and firefighting exercise scenarios
- a ten-bay vehicle storage and equipment maintenance building
- a six-bay fire hall, which includes vehicle response bays, an alarm room, an observation tower and several classrooms
- a building for servicing and maintaining self-contained breathing apparatus (SCBA)
- an automatic fire-suppression systems and fire-detection systems building
- a fire-suppression materials storage building
- a HAZMAT training building
- a storage building for pressurized cylinders
- a fuel holding and dispensing system
- two, two-storey buildings for conducting training on live fires in buildings
- a three-storey, smoke, ladder and rescue training tower

---

12   CFFA Annual Historical Report, 1986.
13   CFFA Annual Historical Report, 1993.

- a confined-space maze to contain artificial smoke for the practice of working in hostile conditions while wearing SCBA
- flammable liquid training pads, complete with mid-size transport aircraft mock-up live-fire flammable liquid training pads, containing mock-ups of a CF5 and a CF18 fighter type aircraft as well as a Twin Huey rotary-wing aircraft
- a roof designed for ventilation practices using rescue saws and chain saws
- a HAZMAT training facility, including a loading dock, a transport truck tank farm, and railroad cars on trackage
- indoor/outdoor auto extrication training areas (cars, highway barriers, light standards, etc.)
- computer-based distance learning training programs utilizing Web CT

Keeping in Step with the Times

A major change occurred in teaching the exercise of command and the management of firefighting and rescue operations, where for years the Fire Service had followed the sound teachings of Lloyd Layman.[14] Nevertheless, when a more advanced and widely accepted Command and Control Program[15] became available, it was introduced into the Fire Service by the CFFA and by individual fire departments. A new fire and emergency training simulator was also developed to teach the practical application of fire ground strategy and tactics in the classroom. This huge indoor training simulator, consisting of a complete simulated military base, provides a means for conducting exercises with video, sound, radio-net and smoke. This was the only simulator of its kind in Canada.[16]

In 1999, technology within CFFA was progressing with the introduction of distance learning, which allows students, using computers, to undergo theoretical training from their home base. Similarly, CFFA made significant strides in the use of thermal-imaging cameras, which proved to be a great tool for safety reasons and very useful for training purposes. This camera permits advanced courses to experience using it in hostile conditions as well as provides instructors with the means of ensuring that the area is left in a safe condition at the end of a training scenario.

Familiarization Training—New Military Engineer Officers

Beginning in 1994, the military engineer officers have attended a four-day familiarization experience whereby they are provided an opportunity to observe what is taught to firefighters by the CFFA as a means of gaining a limited picture of the knowledge and skills firefighters require. They are also provided with a detailed briefing on the firefighter trade.

In 1998, three airfield engineer officers were trained in the basic requirement to become fire officers. This program was commonly referred to as the Cross-Over Program. Three officers were selected from those who volunteered, and each was provided with a study package and given a threshold exam. After passing both practical and theoretical evaluation they received the supervisory training course. All the students were graduated, with two receiving postings to fire departments. Captain Marie Pascal-Rousseau was the first female military fire chief and was stationed at CFB Bagotville.

International Recognition—The CFFA Teaching Institution

Research into the feasibility of accreditation for the CF Fire Service began in 1981 with the CFFM attempting

---

14 *Attacking and Extinguishing Interior Fires*, Lloyd Layman. Published by NFPA, 1995.
15 NFPA 1561—Standard on Fire Department Incident Management System.
16 CFFA Annual Historical Report, 1993.

to have provinces certify courses taught at the Fire School. In 1993, after years of hard work led by Chief Petty Officer First Class John Daley, the IFSAC site-visiting team was invited to CFFA to evaluate the curriculum, the content of individual courses, the teaching staff, and the Academy's facilities. After due diligence, CFFA received accreditation to certify training conducted to satisfy the requirements of twenty NFPA Standards.

Subsequent visits and reviews by IFSAC over the years have resulted in accreditation to certify 23 of a possible 27 levels of firefighter professional qualifications. The CFFA now certifies the following levels: Firefighter I and II; Driver/Operator, Aerial, Pumper and ARFF vehicles; Airport Firefighter; Fire Officer I, II, III and IV; Hazardous Materials Awareness; Hazardous Materials Operations; Hazardous Materials Technician; Hazardous Materials Incident Commander; Fire Inspector I, II and III; Fire Investigation and Public Fire and Life Safety Educator.[17]

Hazardous materials (HAZMAT) qualifications are divided into four levels. The first two are first responder levels taught to all firefighters (Awareness and Operational). The second two are specialist courses to train the advance team to handle the incident (Technician and Commander).

Certification of Individual Qualifications

A committee established by the CFFM set down guidelines and eligibility criteria for certification of firefighters who had completed their training prior to the accreditation of the Academy. In this regard, the Academy has received, reviewed and staffed some 2,000 firefighter accreditation applications. Only those Fire Service members who had earlier been successful graduates of the Academy are eligible to apply for certification.

The CFFA must be re-examined by IFSAC every five years to maintain its accreditation status.

Training Foreign Students

Some interesting additions to the student population were made in October 1986 when, as a result of an External Affairs initiative, a pilot course was conducted for students from the Caribbean. This initiative was referred to as the Caribbean Airports Project. This course was followed by two more courses. In addition, a British firm sent firefighters from Bermuda to the CFFA for training.

CFFA Marketing

In 1989, DND instituted the Delegation of Authority and Accountability Trial (DelegAAT) program which allowed local commanders to begin providing services, at an agreed-upon fee, to non-DND agencies whenever practicable. Initially, the CFFA took advantage of this opportunity by offering full courses, such as Apprentice, Journeyman, Structural, Aircraft Rescue and Firefighting, Hazardous Material Technician and Commander. Some time later, the CFFA began offering seats on under-subscribed courses, conducting weekend and evening seminars, and renting out training facilities. Some of the civilian agencies that have rented CFFA facilities have been the Toronto Fire Department, Honda Canada, Canadian Rangers, Serco Limited, Seneca College, the Ontario Power Generation Corporation, and the Ontario Provincial Police.

The Academy in Profile

The Academy provides classroom teaching and live-operations instruction on a wide range of subjects that span virtually all areas of fire protection. Instructions begin with apprenticeship training, continuing through to senior manager/commander development, with side trips of a specialist nature, to cover fire prevention, fire investigation and hazardous materials. The cumulative

---

[17] NFPA Standard 1035—Public Fire and Life Safety Educator, Level I. Awarded on successful completion of the Fire Prevention and Life Safety course, the Basic Instructional Techniques course, and is qualified to the journeyman level.

number of training days for firefighters, who progress through all courses, excluding re-certification, is 270. This comprehensive program of professional development provides all three military services with a highly qualified Fire Service.

The Apprentice Course

The apprentice course instructs students on life-safety practices on the fire ground, operating fire apparatus, using auxiliary equipment, using and maintaining personal protection equipment, and using and maintaining SCBA. The instructions include structural firefighting, ventilation, salvage and overhaul, forcible entry, fire streams, aircraft rescue and firefighting, hazardous materials to the awareness level,[18] and operating communications equipment. The apprentice course is 65 training days for firefighters who do not have aircraft rescue and firefighting responsibilities, and 76 training days for those that do. This is followed by a period of on-job training.

On-Job Training

On-Job Training Qualification Standard (OJTQS) is utilized at the apprentice level for both the military and civilian firefighters. The training is conducted by fire departments at units employing graduates of the Firefighter Apprentice Course. The OJT program is normally completed within 30 months, and covers harbour shipboard operations, alarm room duties, life-saving duties, aircraft rescue, aircraft arrestor-gear operations, water/ice rescue, driving fire apparatus, and recharge breathing air cylinders.

Completion of the training requires the apprentice to have demonstrated the knowledge and skills necessary to function safely and effectively as an integral member of a firefighting team to achieve a Firefighter I and HAZMAT Awareness level qualification.[19]

The Journeyman Course

The journeyman course is a combination of theory and practical training to take persons who qualified as Firefighter I to Firefighter II,[20] first responder level,[21] HAZMAT First Responder,[22] Vehicle Driver/Operator,[23] and Airport Firefighter.[24] The student expands the depth of knowledge gained during the apprenticeship program, and, performing as an integral part of a firefighting team, exercises and upgrades the skills learned in training and on actual incidents. This course encompasses over 20 broad subject areas and spans 53 training days for firefighters who do not have aircraft rescue and firefighting responsibilities, and 66 training days for those who do.

The Supervisor/Commander Course

This course instructs students in such subjects as command and control for rescue operations, structural firefighting, aircraft rescue and firefighting, water-flow requirements, administration, and safety during emergency operations. The supervisor course spans 45 training days for firefighters who do not have aircraft rescue and firefighting responsibilities and 52 training days for those who do. On completion, the students are qualified to the Fire Officer I level.[25]

The Manager/Commander Course

This course prepares the candidate for the fire chief and deputy fire chief, teaching such skills as middle management and business planning and incident command for natural or man-made disasters, such as

---

18 NFPA 472 Standard for Professional Competence of Responders to Hazardous Materials Incidents.
19 NFPA 1001 Standard for Firefighter Professional Qualifications.
20 Ibid., p. 5 and 11.
21 Ibid., p. 6.
22 NFPA 472, Standard for Professional Competence of Responders to Hazardous Materials Incidents.
23 NFPA 1002, Standard for Fire Department Vehicle Driver/Operator Professional Qualifications.
24 NFPA 1003, Standard for Airport Firefighter Professional Qualifications.
25 NFPA 1021, Standard for Fire Officer Professional Qualifications.

dealing with weapons of mass destruction. The course spans 37 training days. On completion, the students are qualified to the Fire Officer II and III level.[26]

The Fire Investigation Course
At the time of writing, a course training plan had been completed for this course; however, the resources needed to run the course were not at hand. Personnel have been trained in this field by other teaching organizations.

The Fire Prevention and Life Safety Course
This course is to prepare personnel to perform the duties of a fire prevention and life safety inspector with special knowledge and skills to effectively apply the three "Es" of fire-loss prevention: fire loss-limiting engineering, fire safety education, and enforcement of fire safety regulations. Students must be qualified to the Firefighter II level or above and, as a minimum, be of master corporal rank or civilian equivalent. The course spans 37 training days.

The Hazardous Materials Technician Course
This course moved from Lampton College in Sarnia, Ontario, to CFFA in the late 1990s. It is a combination of theoretical and practical training based on NFPA 472[27] that teaches students the skills to enable them to work as a team in a hostile environment. The students perform their tasks in protective gear, which fully encapsulates the wearer, to positively separate them from the surrounding environment. The student is taught how to evaluate incidents, to develop and implement plans, and to utilize the information provided by the HAZMAT placard identification system. The course spans ten training days.

The Hazardous Materials Specialist / Incident Commander Course[28]
This course covers incident command procedures for all HAZMAT responses at any level from minor incidents to those worthy of national attention. As a prerequisite for attendance, students must be qualified to the HAZMAT Technician Level. The training is for platoon chiefs, fire chiefs and other Joint Nuclear Biological and Chemical (JNBC) team members and for any other person who has a need to know. This training provides a reassuring degree of confidence that correct measures will be taken when responding to a hazardous materials incident. The CFFA has also trained persons from jurisdictions external to the military or federal government, such as the City of Toronto and other fire departments. The course spans ten days.

Professional Development of Commissioned Fire Officers, 1952–69
The overarching aim of the basic training provided for newly commissioned officers was to instil the broad range of expectations the military services had for commissioned officers. The six-week course was conducted at RCAF Station Centralia, Ontario. The instructors were exclusively commissioned officers. Some of the key subjects in the curriculum were general service knowledge, written correspondence, both internal and external; personnel management; and social interaction, as a means of preparing the students to execute the more senior management responsibilities they might expect to encounter as a commissioned officer.

Professional Development of Commissioned Fire Officers, 1970 Onwards
A major change in policy for training newly commissioned fire officers was put into effect in 1970,

---

26 Ibid., p. 9 and 11.
27 NFPA 472, Standard for Professional Competence of Responders to Hazardous Materials Incidents.
28 As a prerequisite for attendance, students must be qualified to the HAZMAT Technician Level.

when the focus was directed more towards the military engineer branch of the service than the military services as a whole. They began to be trained as military engineers through attendance on the basic military engineer officer course of approximately one year in duration, administered by the Canadian Forces School of Military Engineering. The aim was that every fire officer be fully trained in the military engineering skills required in a field squadron or construction engineering section. For many years this course was taught at CFB Chilliwack, British Columbia, before moving to CFB Gagetown, New Brunswick. Course content included such subjects as parade ground drill, field craft, small arms, physical fitness training, map reading, first aid, rigging, compressors, field defences, fixed equipment bridges, watermanship, equipment bridging, mine warfare, water supply, construction engineering, and military roads and grounds. The only item relating to the Fire Service was a familiarity visit to the Fire School/Academy.

Professional Development of Commissioned Fire Officers, Fire Service Specific

Some training, specific to the professional development of commissioned fire officers, such as an on-scene commander course, an automatic fire-protection systems course, a fire investigation course, other courses taught at the CFFA, and peripheral training, such as Staff College and language training, has been provided.

Summary, Professional Development of Commissioned Fire Officers

Commissioned fire officers have had, in the main, to draw from past experience in the Fire Service and on-job training, to achieve the desired level of proficiency in their new role. Generally, this has usually been as fire chief, for approximately three to nine years, sometimes interspersed with appointments as staff officers in headquarters formations.

Other than the staff officer positions, this is the same path travelled by NCOs appointed to the position of fire chief. The unchallengeable fact is that, irrespective of what route is followed, you are never fire chief until you are, well, fire chief. Overall, it can be argued that time spent on training on non-Fire Service matters retards, rather than accelerates, the professional development of FSOs.

Recertification Training

The CFFA has been responsible for re-certifying all firefighters who are required to serve as ARFFV driver-operators to refresh their aircraft rescue and firefighting skills in accordance with a Standard NATO Agreement[29] and as recommended by the NFPA.[30] The certification training spans four training days.

Distance Learning

In 1998, in response to cutbacks in Academy staff at a time when training requirements were increasing, alternatives were developed to assist in accomplishing the mission. The other important factor was that attendance of firefighters on course for long periods created difficulties for fire chiefs. A potential solution was to take advantage of the available computer technologies, which could allow students to complete theoretical training at their home bases. Training in firefighting operations, primarily, remained the responsibility of the CFFA.

The essentials of the concept was that a student would register and receive a log-in identification code. Students could then proceed at their own pace to complete the training modules of the course. Questions or problems on the course are sent to CFFA for

---

29 Stanag No. 7145, Minimum Core Competency Levels and Proficiency Skills for NATO Firefighters.
30 NFPA 405, Recommended Practice for the Recurring Proficiency Training of Aircraft Rescue and Fire-Fighting Services, 1999 Edition.

answers from assigned instructors. Once the modules are completed, arrangements are made for the student to write the exam through the local fire training officer to ensure the correct student attends alone to complete the exam without any assistance. The computerized program includes a time limit within which the exam must be completed. The exam is marked and recorded at the CFFA. Following successful completion of the exam, the student is then eligible to attend a course at the CFFA to complete whatever training is necessary to achieve the desired qualification.

In order to evaluate the concept, Sergeant Brian Ross took on the task of developing a test course. The Supervisor's Course was chosen for the test due to its limited amount of theoretical training. Given the time frame available, it was decided that the course would be run as usual with the exception that during the first two weeks at Borden, the students would be required to teach themselves the theoretical portion and an instructor would be available to answer questions or solve problems with the initial package. At the end of the two weeks, the students wrote an exam and then commenced the portions of theoretical training that would continue to be conducted in the CFFA classrooms. This initial trial demonstrated that distance learning had the potential to be an excellent teaching tool.

A Film and a Story
Three Minutes to Live, an aircraft rescue and firefighting training film, the production of which was an outstanding achievement and an interesting story. The story, from the memoirs of Major Phil Brown, is recorded below.

> I was assigned to give a three-hour lecture to the Field Engineering course at CFB Chilliwack around 1961. I worked up a slide series and thought what we really needed was a film that showed the technical aspects of our trade. One thing led to another (as it usually does) and the idea of a standard, purely technical film would be a decided asset, not only to our crash-rescue training at Borden, but would also be a good PR film anytime we were asked to give a lecture to civilian fire departments who wanted to know why we claimed to be special because of our triple role. If you don't stick your neck out you'll never get ahead. I put in a request to the Director of Training Aids. Next thing I knew, it received preliminary approval subject to a satisfactory script. I stuck to it being a purely technical film and drew up a script.
>
> About this time I was posted to Camp Borden as OC FFS. I completely forgot about the proposed film until a representative from the National Film Board and a corporal from the RCAF Photo Establishment turned up to discuss the proposed film and look at our facilities. The corporal insisted that it was beyond their scope, so a civilian photographer was suggested. To cut a long story short, a crew turned up a few weeks later. As I insisted filming must not interfere with the training program, it took a fair bit of scheduling.
>
> I assigned Flight Sergeant Cliff Brooks as my liaison representative, and couldn't have made a better choice. Cliff selected a team of instructors who could be spared for a couple of weeks and we held back the graduating course to do the actual firefighting. For realism, we imported a stored CF-100 and went at it.

A bit of humour crept in when Cliff wanted a live body, for realism, to be rescued. At a TGIF in the mess, I collared a group of pilots from Ferry Flight and asked if anyone wanted to be a movie actor. I played down the role until one volunteered and we were in business.

Shooting took about a month and didn't interfere with the training schedule, thanks to Cliff Brooks. When the crew packed up and left, there was the usual after shooting reviews, which resulted in Cliff and me making several trips to the Montreal National Film Board Headquarters. When we were satisfied, they informed us we could start using it.

It came as a complete surprise a few months later when we got a message from the National Committee on Films for Safety advising us our film had won the highest award in the Occupational Films for 1964, and the RCAF was asked to send a representative to accept the award at their annual presentation ceremonies. The powers that be graciously asked me and my wife to represent the RCAF and go to Chicago and accept the award. We accepted, of course, and the rest is history. I successfully fought off an AFHQ request to send the plaque to them by insisting it was made at the Fire School and should stay there, where I hope it still resides.

Copies of the film were bought from the NFB by several countries that had Fire Schools, and we received several letters of congratulations from them.

This was not an individual award, but included a team of Instructors from the Fire School headed by Flight Sergeant Cliff Brooks.

The commemorative plaque honouring the training film, *Three Minutes to Live*.

Events of Special Interest

1976
The first two female military firefighters graduated from QL-3 course 7603 into the Fire Service. Private Sherry Rickard posted to CFB North Bay and Private Shannon McLaughlin posted to CFB Comox.

1985
On May 31, 1985, a tornado struck the city of Barrie, Ontario, killing 11 people and injuring hundreds of others in a demonstration of the awesome power of nature. Over 1,000 buildings were either destroyed or damaged. All staff and students responded to assist the local fire departments in search and rescue operations.

1990
In 1990, the CFFA was host to the 50th anniversary of the Fire Service. The celebrations included a formal parade, a meet and greet function, a live-fire demonstration, and other events. Firefighters came from all across Canada to attend this event.

1998
In addition to their academic duties, in January–February 1998, CFFA staff responded to a national emergency in Quebec. This need for help was due to the effects of a severe ice storm in Eastern Ontario and Quebec that severely crippled the power grids in each province. The initial tasking for 15 CFFA instructors was to set up a HAZMAT response team in St-Jean D'Iberville, Quebec, utilizing the HAZMAT trailer from St-Hubert. In addition, the CFFA team, led by Chief Warrant Officer Yvon Serre, engaged themselves in augmenting the St-Jean D'Iberville Fire Department, which was running near exhaustion. The first task was to take control of a major fire at a strip-mall fire that the St-Jean D'Iberville firefighters had been battling for some time. The team took up residency in the St-Jean D'Iberville fire hall and was employed as first responders on approximately 284 emergency calls during a ten-day period. All team members received personal appreciation certificates from the mayor of St-Jean D'Iberville and the premier of Quebec, Lucien Bouchard.

1999
The year of 1999 was a medium-paced year for the CFFA, with a light load of students to train and courses to run. This was due in part to all Priority 5 training and below being cancelled from October 1999 to March 2000. Further, DND was fully involved with Operation ABACUS, a contingency preparation for potential Year 2000 computer problems.

Summary
The Academy's role, even though it evolves to meet adjusted requirements and technological advances, remains very much the same as it was in the beginning. The CFFA mandate continues to conduct individual training in all subject areas where the Fire Service bears a responsibility, including fire-safety education of military personnel and civilian employees of DND, fire loss-limiting engineering, enforcement of fire codes, regulations and orders, the operating principles of installed fire protection systems, structural firefighting operations, aircraft rescue and firefighting operations, HAZMAT incidents emergency response and on-scene emergency site management, and situation control.

Over the years the Fire Service and, in the broader sense, the military services have been well served by the various fire schools and finally by the Fire Academy. These institutions have played an important role in preparing Fire Service personnel to deal with a wide range of tasks in a high-risk environment with minimal backup, in the process of protecting critically important resources from fire.

Picture Gallery VI

Basic Firefighter Course No. 1, May 22, 1941, Trenton. This was the first class to graduate from the Fire School at RCAF Station Trenton. Back Row: NR.
Front Row: Sgt Jones, NR, F/O Fulford, S/L Massey, NR and Sgt Matthews.

Basic Firefighter Course, 1951, Trenton.
Back Row: AC Porter, AC Harley, AC Baptiste and LAC MacSepheny.
Front Row: AC Walker, AC Hendry and AC Waggie.

Basic Course 5201, 1952, Aylmer.
Back Row: AC Dolan, AC Roth, AC Outhouse, LAC Thomas, AC MacLean, AC Peterson, AC Falkner and AC Morden.
Centre Row: Cpl MacDonald, AC Leeman, AC Henry, AC Clarke, AC Roop, AC Berube, AC Chamberlain and Cpl Collins.
Front Row: AC Ponton, AC Allen, AC Craig, WO1 Lefebvre, F/S Hollins, AC MacDonald, AC Parker and AC Sisk.

Basic Firefighter Course, 1951, Trenton.
Back Row: AC Porter, AC Harley, AC Baptiste and LAC MacSepheny.
Front Row: AC Walker, AC Hendry and AC Waggie.

CHAPTER 6 • PROFESSIONAL DEVELOPMENT

Basic Course 144, 1954.
Back Row: AC Maillet, AC McKillop, AC Carmichael, AC Brough, AC Baginski and AC Bowers.
Front Row: AC McInnis, AC Phil Thormin, Sgt Ken Stokes, WO2 Bob Edwards, Sgt Phil Brown, Sgt Bill MacDonald and AC Feather.

Supervisors Course, 1957.
Back Row: Carl Brady, NR, NR, NR, Frank Fertich, NR and Cliff Brooks.
Front Row: Doug Stevenson, NR, Ken Wolstenholme, Bert Deveau, Don Curtis and NR.

Basic Course 130, 1961.
Back Row: AC Al Groulx, AC Bruce Jones, AC Ben Burgess, AC Chuck Roberson, AC Neil MacFarland. AC Rick Other, AC Bob Thorne, AC Danny Summers and AC Reg Tressel.
Front Row: AC Bob Cole, AC Joe Carty, AC Arnie Dauphinee, AC Gil Verville, AC Roy Wilson, AC Pete Raycove, AC Norm Bournival and Junior Kirkum.

Fire School staff, 1964.
Back Row: LAC Lebeau, Cpl Ellison, Cpl Ronan, Cpl Whitehead, Cpl Kapralik, Cpl Thompson, Cpl Breckenridge, Cpl Buck, Cpl Morrison, Cpl Orr, Cpl Anderson and Cpl Smith.
Centre Row: Cpl Blanchette, Cpl Comeau, Cpl Foucault, Cpl Keon, Cpl Gobiel, LAC Henderson, Cpl Christenson, Cpl Holley, Cpl Pritchard, Sgt Taggart, Sgt Porter, Sgt Briggs and Cpl Gregson.
Front Row: F/S Dyck, F/S Tharp, F/S Teed, WO2 Morash, F/L Brown, Mrs. Robson, F/S Hollins, F/S Brooks and F/S Gagne.

Automatic Fire Suppression and Detection Systems Course 6401.
Back Row: F/S Teed, Cpl Pritchard, Cpl Comeau, F/S Brooks, Cpl Phillips, Cpl Gobeil, Cpl Orr and Sgt McGregor.
Front Row: Sgt Briggs, Cpl Robichaud, F/S Gillespie, F/S Tharp, F/S Dyke, F/S Gagne and Sgt Porter.

Automatic Fire Suppression and Detection Systems Course 6402.
Back Row: Sgt Brown, F/S Cottrell, Sgt Lockhart, Sgt Swihart, Sgt Ebel and NR.
Centre Row: Sgt Brown, Sgt Finnigan, Sgt Freshwater, Sgt Gilbert, Cpl Smith and Mr. Lay.
Front Row: Sgt Porter, F/S Tharp, F/L Brown, F/S Teed, Sgt Briggs and Sgt McGregor.

Automatic Fire Suppression and Detection Systems Course 6403.
Back Row: Sgt Kroeplin, Sgt Dykes, WO2 Wood, Sgt MacDonald, Mr. Comeau, Sgt Johnston, Mr. Scanlan, Sgt Green, Sgt Dundas and Sgt Starchuk.
Front Row: F/S McEvoy, Sgt Porter, F/S Teed, F/L Brown, Sgt Briggs, Sgt McGregor and Sgt Doyle.

Automatic Fire Suppression and Detection Systems Course 6404.
Back Row: Cpl Kapralik, Sgt Fowler, Cpl Gregson, Cpl Franc-Guimond, Cpl Landroche and Sgt Merrill.
Centre Row: Sgt Walker, Sgt Jack, Sgt Zillman, Sgt Faulkner, Sgt Hotson and Sgt Bookham.
Front Row: Sgt Porter, Sgt McGregor, Sgt Briggs, F/S Teed, F/L Brown, F/S McEvoy and Sgt Briggs.

Automatic Fire Suppression and Detection Systems Course 6604.
Back Row: Sgt Saunders, Mr. Schneider, Sgt Gallant, Sgt McCullough, Mr. Fanslav and Sgt Monck.
Front Row: Sgt Briggs, F/S Teed, F/L Brown, F/S McEvoy and Sgt Hebert.

Automatic Fire Suppression and Detection Systems Course 6606.
Back Row: Cpl Hansford, Cpl O'Meara, Cpl Lynch, Cpl Bedell, Cpl Cranidge, Cpl Livingston, Cpl Savoury, Cpl Copeland, Cpl Dewar and Cpl Lobban.
Front Row: Cpl Williamson, Sgt Bowie, F/S Teed, F/S Murphy, WO2 Duguid, F/S McEvoy, Sgt Briggs and Sgt Hebert.

Automatic Fire Suppression and Detection Systems Course 6607.
Back Row: Cpl Heudes, Cpl Wildes, Cpl Bedard, Cpl Thomas and S/S McLachlan.
Centre Row: Cpl Lofgren, MCpl Emes, Cpl Nicholson, Cpl Goodman, Sgt Ross, Cpl Jensen and Cpl Herman.

TQ 6A Course 6801.
Back Row: Curry, Taylor, Reid, Bigras, Quick, Morrison, Clark and Swaffield.
Centre Row: Moxam, Denhollander, McBride, Howlett, Coakley, Bruton and Neidzwiecki. Front Row: Boudreau, Horton, McIntosh, Ferris, Gervais, Colbourne and Hoeg.

TQ 6A Course 6901.
Back Row: Bartlett, Singleton, Ronan, Brook, Hunt and Chapman.
Centre Row: Brown, Doherty, Fulmer, Trudeau, Little and St John.
Front Row: St Onge, Rockey Ferris, Murphy, Dale and Porter.

TQ 6B Course 7201.
Back Row: Sgt Keith-Murray, Sgt Daley, Sgt McGallan and Sgt Nicoletti.
Center Row: Sgt Knight, Sgt Pope, WO Rumsam, MWO McIntosh and WO Dark.
Front Row: Sgt Herman, WO Bigras, MWO Rockey and Sgt Hunt.

CHAPTER 6 • PROFESSIONAL DEVELOPMENT

TQ 6A Course 7301.
Back Row: Mr. Smith, MCpl Campbell, Sgt Jarret, Sgt Little, Mr. Totten and Mr. Fleming.
Centre Row: MCpl Embrett, WO Bartlett, WO Ronan, Sgt Hunt, WO Chapman and Mr. Ryan.
Front Row: Cpl Giesbrect, Sgt Herman, Major Haggart, Sgt McGallan and MCpl Nye.

TQ 6A Course 7401.
Back Row: MCpl Potvin, Sgt MacDonald, Sgt Rowat, Sgt Cyr, Sgt Bayliss and Sgt Foley.
Centre Row: Sgt Cote, Sgt Terrio, Sgt Hodges, MCpl Geddes, Sgt Quinn and MCpl Arsenault.
Front Row: Sgt Herman, Major Ferris, WO Wolstenholme and Cpl Giesbrecht.

TQ 6A Course 7501.
Back Row: Sadler, Theriault, Auger, Bonner and Moore.
Centre Row: Wright, Willis, Hills, Nadeau and Benn.
Front Row: Lofgren, Herman, Ferris, Wolstenholme and Nye.

Pte Shannon McLaughlin and Pte Sherry Rickard, the first females to complete the firefighter basic course.

Basic Course 7604.
Back Row: AC Willshire, AC Chambo, AC Paton, AC Donald, AC Gurmin, AC D.M.Tucker, AC Winsor, AC Lamarche and AC Bogart.
Centre Row: AC Richardson, AC McConkey, AC Corbin, AC Hynes, AC S.W. Tucker, AC Richard, AC Nurse and AC Payette.
Front Row: Cpl Olster, WO L'Abbe, Capt Maxwell, Sgt Walker, MCpl Dionne and MCpl Blacquiere.

Basic Course 7701.
Back Row: Weir, Wilson, K.C. Wood, Wollinger, Grant, Smith, Gomes and Moeckl.
Centre Row: Dixon, Johnson, McGowan, Vanpol, W.S. Wood, Cormier, Cade, Freeman, Phelps and Donnelly.
Front Row: Hole, Blacquiere, Walker, Maxwell, Dionne and Jansen.

TQ 6A Course 7701.
Back Row: Young, Carson, Burton, Apblett, Connelly and Olsen.
Centre Row: Gark, Throwsdale, Deion, Allen, Batchelor and Durdle.
Front Row: Fox, Bonner, Rumsam, Maxwell, Keough and Currier.

TQ 6A 7703.
Back Row: Sgt Thompson, Sgt Stewart, Sgt McGloin, Sgt Armstrong, Sgt Ferguson, Sgt McLaren and Sgt Bisson.
Centre Row: Sgt Alexander, Sgt Jones, Sgt Whitehead, Sgt Jay and Sgt Neufeld.
Front Row: MCpl Langlois, Sgt Bonner, WO Keough, WO Benn and Sgt Hole.

TQ 6A Course 7802.
Back Row: Sgt Farrell, Sgt Fox, Sgt Wilson, Sgt Ranford, Sgt Hunter and Sgt Cestnick.
Centre Row: Sgt Botting, Sgt D. Brown, Sgt Ennis, Sgt Deans, Sgt P. Brown and Sgt Galaugher.
Front Row: Sgt Bonner, MWO Rumsam, Lt Gordon, Sgt Hole and Sgt Oltsher.

TQ 6A 7803.
Back Row: Sgt Kurliak, Sgt Simmons, MCpl Chellew, MCpl Forester, Sgt Zandewr, MCpl Olynek, Sgt Smith, Sgt McDonnell, Sgt Daniels, Sgt Gray and Sgt Woodford.
Front Row: Sgt Oltsher, MWO Rumsam, Lt Gordon, WO Bonner and Sgt Hole.

QL6A Course 0401, 2004.
Back Row: MCpl Mario Lafreniere, MCpl Kelly Smith, MCpl Bob Gale and MCpl Robert Comeau.
Front Row: MCpl Noel Green, MCpl Jacques Lacharite, MCpl Marius Leblanc, MCpl Jean Boilard, MCpl Brad Nancekivell and MCpl Rick Pickell.

CHAPTER VII

FIRE PREVENTION

GENERAL

Philosophy and Structure

Historically, Canada's military services have always recognized the need for comprehensive fire prevention programs as an effective means for protecting life and property against the ravages of fire. This attention to fire prevention begins with the inclusion of fire and life safety features in structures during the design phase through to the application of fire-engineering design principles. This approach is continued by subjecting conditions of occupancy and the practices and procedures used in operations to ongoing scrutiny, and by endeavouring to ensure that the military and civilian work are cognizant of the risk posed by fire.

At headquarters formations fire-prevention issues have been given a great deal of attention by the staff of the CFFM's office and by the staff of each Command Fire Marshal's office to:

- review and assess hazards relative to materials or practices;
- develop policy to alleviate the risks posed by recognized hazards;
- keep individual units informed;
- evaluate fire investigation reports; and
- monitor the enforcement of fire-safety policy.

In support of this, over the years, many members of the Fire Service have served as members of committees developing fire-related codes and standards in Canada and the United States. At unit level, given sufficient staffing, each base or establishment generally has had staff specifically dedicated to fire prevention. At larger units, fire chiefs have set up fire prevention bureaus under the command of a chief fire inspector with a number of fire inspectors to carry out the various programs. In addition, all have, to varying degrees, been consistent in their efforts to prevent unwanted fires.

Raising the Quality Standard
The most efficient method of fighting fires is to prevent them before they start. The example set by the army, which employed specially trained fire inspectors in the early years, did not go unnoticed. As an illustration of this, just over two years after unification, the Fire Service had developed a fire inspector course, which was introduced into the curriculum of the Fire School. More recently this course has been taught at CFFA since its inception and is certified by IFSAC. The course syllabi cover all areas of fire loss-limiting engineering, inspection of facilities and the promotion of fire safety. In addition to the fire inspector course, a fire engineering course was established. This policy allowed up to two firefighters a year to attend a university fire engineering course under the University Training Plan. Graduates were commissioned and provided another source of expertise for fire prevention.

The THREE "Es" of FIRE PREVENTION

Overview
The Fire Service has long recognized the three "Es"—engineering, education and enforcement—as the foundation of fire prevention, recognizing that all fire prevention activities fit under the umbrella of these broad terms. To varying degrees, all fire service organizations have directed their efforts towards applying the principles of fire loss-limiting engineering, promoting an awareness of the need to practice fire safety, and to comply with fire-safety regulations. The simplistic, but not easily achieved, goal has been keyed to the Fire Triangle. Specifically, it is to prevent the dreaded trio of fuel, sufficient oxygen to support combustion, and sufficient heat to cause ignition, from being in the same place at the same time.

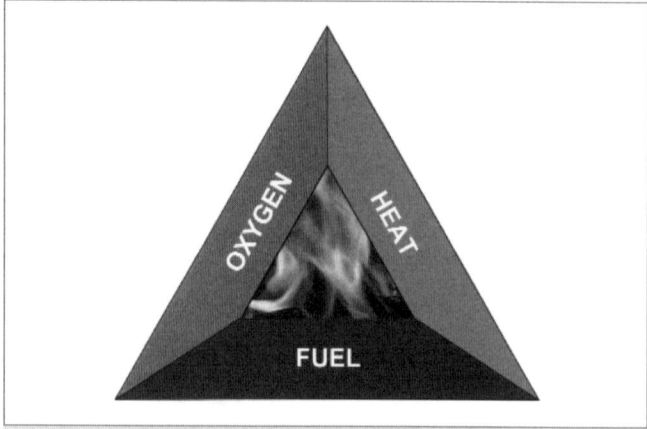

The "Terrible Trio."

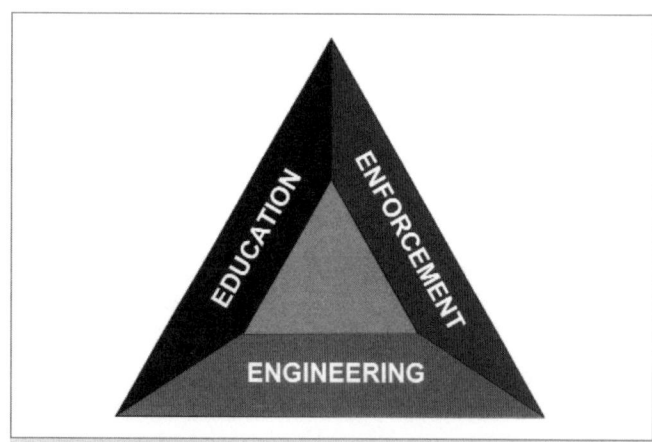

The "Tenacious Troika."

Fire loss-limiting engineering has included the review of drawings and specifications for new construction and renovations to ensure buildings are designed and sited to meet or exceed current building code standards. Fire prevention education covers lectures, demonstrations, fire department tours and training programs for a wide range of personnel and their dependants. The highlight of the fire prevention year is Fire Prevention Week, which occurs in October, and involves the departments in many activities with the general public. The enforcement role has been executed through the routine inspection of buildings and facilities by Fire Service personnel and by special inspections for operational or social events.

For many years the bases and stations entered a summary of their fire prevention activities in photograph album format and submitted it as an entry in NFPA's annual fire prevention contest. Command fire marshals and staff from the office of the CFFM judged the entries, after which they were forwarded to Boston for final judging by NFPA. The quality of the fire department's fire prevention programs is reflected in the many grand awards and honorary mentions that hang on the walls of their organizations. In later years, a more statistical and action-oriented format was developed and used.

The lead unit in a Fire Prevention Week Parade at RCAF Station Uplands (Ottawa) in 1964. The driver is a local farmer. Riding "shotgun" in a sliver bunker suit is Master Roy MacLean, and fuelling the boiler is LAC George Gray.

FIRE LOSS-LIMITING ENGINEERING

The Beginnings

Construction of several large buildings was undertaken at RCAF Station Trenton during the 1930s as part of a Depression-era, government make-work program. While non-combustible material, such as poured concrete and steel, was extensively used, the structures incorporated sprinkler systems throughout. Design authorities of that time were effectively constructing buildings according to sound fire-protection principles. This was in marked contrast to the situation during the Second World War, when the haste to build wartime facilities took precedence over fire protection. For example, many buildings generally lacked a sprinkler system, even though they were fabricated from highly combustible material. Most of those buildings were considered temporary and therefore not worthy of the added expense, time and material needed to build them in accordance with the building and fire codes of the period. While some buildings did incorporate fire-safety features of sorts, these were mainly design considerations for special occupancies, such as fuel tender garages. Methods of venting hangar roofs and installing fire doors, where separation of workshops was desirable, received special consideration.[1]

Several bases installed total flood carbon-dioxide systems in flammable-liquid storage spaces and, to some extent, dry-chemical and carbon-dioxide systems provided protection to areas, such as exhaust hoods over mess hall cooking facilities. These extinguishing systems were later replaced with systems referred to as wet chemical. Wet-chemical systems were relatively clean agents and were easier and less expensive to maintain. As well, unlike dry-chemical systems, which required a lot of clean up after use, or CO_2 which, in confined spaces, posed a health hazard, the wet chemical avoided these shortcomings.

---

[1] National Archives file: 3494, 866-20-90.

Evaluating Construction Projects—
A Structured Method[2]

At Air Command in the mid 1970s, the Command Fire Marshal was responsible for reviewing all of the fire engineering requirements for all drawings and specifications for new construction and maintenance projects. There was a high volume of work in this important field, and the need for thoroughness and accuracy was paramount. As a means of satisfying this requirement, the CFM and his staff developed fire engineering *pro forma* check lists, which provided a step-by-step reference to life safety and fire engineering, which eventually became the standard for engineering project review throughout DND. This staff work resulted in raising the profile and professionalism of Fire Service personnel, as the Military Engineer Branch acknowledged the fire engineering expertise provided to project engineers and architects.

The Origin of Sprinkler Systems[3]

Sprinklers were invented in the US about 1870 to meet the need for instant, automatic suppression of fires in certain types of industrial buildings, notably cotton mills. Rules and regulations for their installation were published under the auspices of the NFPA in the late 1800s. Known for many years as NFPA 13, it was the first standard published by NFPA, which now lists approximately 300.

For a long time, Canada's military services did not consider installation of fixed fire-suppression systems in their properties.[4] Reliance against fire loss in properties was placed on structural integrity, fire-resistance rating of structural assemblies, firewalls, fire doors, vertical fire separations, and a well-trained, efficient fire department summoned by a well-maintained fire detection and alarm system. The Navy did have a few large unheated warehouse-type buildings for storage of high-piled material that were protected by dry-pipe automatic sprinkler systems.

Automatic Sprinkler Systems

Automatic sprinkler systems have been the foremost means of protecting valuable assets. Due to the outstanding reliability and effectiveness of wet-pipe sprinkler systems, they were the systems of choice wherever conditions would permit. For many years, the design of sprinkler systems was based on using the pipe schedule method. In later years, the requirement concepts remained the same, but the systems were hydraulically designed. This design method considered available water supplies relative to the amount of water required for particular locations and areas, and was quite cost-effective. Areas subject to freezing were provided with dry-pipe sprinkler systems, while locations that were sensitive to water damage would be protected with pre-action sprinkler systems.

Aircraft hangars were traditionally protected with wet-pipe sprinkler systems; however, new extinguishing agents made it possible to provide protection for the flammable and combustible fuels contained in the housed aircraft. Aqueous film-forming foam, deluge sprinkler systems became the preferred protection for new hangars. These were hydraulically designed, using regular, open sprinkler heads, and for most installations, fire pumps had to be designed into the system to ensure that the calculated quantities of water and pressures would be provided.

Foam-Deluge Systems in Aircraft Hangars[5]

In the mid 1950s, the Air Force obtained sufficient funds to construct a few greatly needed large aircraft hangars; the design was done by outside consultants.

---

2   Memoirs of Lieutenant Colonel H. Singleton (1960–93) (unpublished).
3   NFPA 13, Installation of Sprinklers.
4   Memoirs of Mr. Kurt Brewer (1955–84), Specialist, Fire Suppression Systems, 2003 (unpublished).
5   Ibid., p. 3.

As an afterthought, two of these aircraft hangars were equipped with a foam sprinkler system which was both expensive and complicated. The foam-deluge sprinkler system had not been part of the original design of the building, and it was necessary to construct a large pump house outside the building for the installation of large centrifugal fire pumps, creating a maintenance problem. The installation of a fixed foam deluge sprinkler system was a first.

The government was so proud of this installation that it invited a number of foreign military attachés to attend the acceptance test at CFB Greenwood, Nova Scotia. The flight from Ottawa to Greenwood was in a propeller-driven aircraft without any sound-proofing barriers, and the engine noise was deafening. Upon arrival in Greenwood, the group was split in two and loaded into two buses. It happened that the Soviet attaché, who barely understood English, was in one bus and the other attachés were in another bus. Out of the blue, someone in authority decided that the Soviet should not be allowed to see the hangar. The bus driver was instructed to take his bus the long way through the base, including all streets in the married quarters, and to end up at the officers' mess for lunch.

Experience showed that foam-deluge systems were not cost-effective and were prone to problems, such as accidental discharge and the breaking of the high-pressure underground piping. Thereafter, installations had to be tried, tested, and approved by use in industry before implementation by DND. In the early 1960s, the department started installing wet-pipe automatic sprinkler systems in wood-construction wartime aircraft hangars. The design was done by DND in accordance with the standards of the NFPA. However, the department was instructed to have the finished sprinkler drawings approved by ULC before being passed to Defence Construction Canada for action. To become independent of ULC, 19 individuals from DND were sent to the Engineering Branch of Factory Mutual Research Corporation to participate in a three-week fire-protection engineering, field engineer course. Sadly, only one member of this group would use the knowledge gained on behalf of DND. In time, a great number of buildings were equipped with wet-pipe automatic sprinkler systems. Adding aqueous film-forming foam to some automatic water sprinkler systems improved its capability of extinguishing flammable liquid fires.

Carbon Dioxide Fixed-Pipe Systems[6]
CO_2 had long been a favourite extinguishing agent, primarily in hand-held units and later in wheeled cylinders. Fixed total flooding CO_2 automatic fire protection systems were used for exhaust hoods on commercial cooking equipment, in liquid fuel pump-rooms, flammable stores and other applications where the discharge of water should be avoided. CO_2 total flooding was also used for the protection of electronic data processing and control systems, until it was elbowed aside by halon. For this type of hazard, a CO_2 concentration of 30% was required.

Halogenated Fire-Extinguishing Agents
A halogenated compound is one that contains one or more atoms of an element of the halogen series: fluorine, chlorine, bromine, and iodine. The numerical system for identifying the various blends of halogenated hydrocarbons was devised by the US Army Corps of Engineers. The first digit in the number represents the number of carbon atoms in the compound molecule; the second digit, the number of fluorine atoms; the third digit, the number of chlorine atoms; the fourth digit, the number of bromine atoms; and the fifth digit, the number of iodine atoms. Terminal zeros are dropped.

---

[6] Ibid., p. 3.

The manner by which these agents extinguish fire is highly unorthodox, in that it does not involve cooling a fuel below its ignition temperature, denying the fire sufficient oxygen to support combustion, or removal of the fuel source. Rather, it is believed that it interrupts the chemical combination attended by production of heat and light that constitutes combustion. The agents are sufficiently unique as to lead them to be identified by the name "Halon" which is exclusive to these compounds and is not used as a unit of language.

There are significant differences in the characteristics of the two agents. Specifically, Halon 1301 is in a gaseous state immediately upon release to atmosphere, and is therefore directionless, while Halon 1211 discharged in liquid form provides a reach of stream in the order of 8 ft to 12 ft / 2.4 m to 3.7 m. For Canada's military, the application was almost exclusively Halon 1211 for hand-portable fire extinguishers and Halon 1301 for fixed-in-place systems.

Bromochlordifluoromethane and bromotrifluoromethane are the chemical descriptions of the extinguishing agents Halon 1301 and Halon 1211, respectively.

Canada's Military Embraces Halon
Fires and explosions are among the greatest threats to the safety of military personnel and the survivability of military aircraft, land combat vehicles, naval ships and military land facilities in both peacetime and during combat operations. To reduce this significant risk, Halon 1301 and Halon 1211 were introduced and became the agents of choice for many NATO militaries shortly after World War II, and its popularity grew significantly during the 1960s. In the mid 1970s, the use of Halon started to make significant inroads with the civilian telecommunication world in North America, as well as with high-tech computer facilities, and by 1980, it became the agent of choice within these industries.

The Canadian military was somewhat slow off the mark to accept this "wonder gas," as it came to be known; the first serious usage of the agent did not occur until the mid to late 1970s. Some fire-suppression units were installed in army combat vehicles in the early 1970s by Douglas Fire Safety Systems under the direction of Doug Stevenson, an ex-military firefighter turned entrepreneur, who was the company's president. These systems became the forerunner for the more advanced systems that were later designed and installed during the army's Armoured Vehicle Modernization Program. It is believed, although not confirmed, that Halon was also used by the air force to protect aircraft engines around this same time. As best as could be determined, these systems were the beginning of DND's relationship with Halon.[7] However, from that point, the interest of the Canadian military escalated rapidly.

Demands for Halon Fire-Suppression Systems
Many organizations within the Canadian Forces enthusiastically embraced the characteristics of Halon and described needs for systems big and small. Henceforth, having taken up this challenge, DND quickly proceeded in the late 1970s and designed and installed these uniquely different fire-suppression systems.

In the early 1980s, DND took on the challenge of designing and installing one of North America's largest total flood systems in the Hornell Centre at CFB Greenwood. This system was to protect the then new Aurora aircraft training modules, including the flight deck and the on-scene mission control centre for maritime surveillance by ships and aircraft. Even though DND had not achieved total success with these systems, they were quickly followed by designing and installing other equally large systems in the new frigate training facility at Halifax Dockyard and in the Air

---

[7] Memoirs of Major Jack Henderson (1961–92) (unpublished, 2004).

Force's Command Centre in its new HQ Building in Winnipeg. Also, during this period, very small systems were installed (under 1 lb / 0.5 kg) in the washroom waste receptacles on Boeing 707 aircraft. Fortunately, soon after the initial installation of these units, an accidental fire occurred in one of the receptacles at or near the "point of no return" during a transatlantic flight. The unit activated and held the fire in check until stewards were able to fully extinguish the fire with the use of hand-portable Halon 1211 fire extinguishers.

As experience was gained, the agent was used to protect risks, such as the CF-18 aircraft engine test facilities in CFB Cold Lake, Alberta, and in the army's armoured combat vehicles. Also the navy had significant requirements for their new high-tech frigate class ships. In fact, each new frigate had 30-plus systems installed prior to becoming operational.

The Halon 1301 fixed-pipe systems, each of which embodied a myriad of electronic fire detection and systems-control devices, were extremely sensitive and totally unforgiving of human errors. Combine this with fact that, due to the newness of the concept, neither consultants, installation contractors nor the authority having jurisdiction had any meaningful experience in dealing with the systems. Mishaps were virtually inevitable and many occurred. The most significant was the accidental release into the atmosphere of approximately 9,000 lb / 4080 kg of Halon in the Hornell Centre. In this incident, not just the primary system that was on-line, but also the reserve back-up system that was supposed and thought to be off-line, was activated. In light of the adverse effect the agent has on the environment, it was a shocking and disturbing event.

Needless to say, DND now had the attention of Environment Canada, and it was decided that an FSO would be appointed as the project manager to oversee the Halon program and related issues and to answer Ministerial Inquiries and, of course, to keep key persons in senior positions informed of the next challenge or conquest involving that "wonder gas" called Halon 1301.

Halon—Waxing and Waning

The advantages of clean extinguishing agents, such as Halon 1301 and Halon 1211, were widely recognized. Halon 1301, being gaseous in nature, was found to be ideal for protecting major computer installations that generally include difficult to reach under-floor spaces or areas that did not have adequate water supplies readily available to support other types of fire-suppression systems. Halon 1211 could be applied directly onto electrical or electronic equipment without causing collateral damage. In fact, Halon 1301 became an extinguishing agent that was widely used in both occupied and unoccupied locations. This trend continued until about the mid 1980s, when it became evident that these products were factors relating to the depletion of the ozone layer. Consequently, the use of Halon 1301 and Halon 1211 was restricted to those risks of critical importance, and steps were taken to obtain more environmentally friendly extinguishing agents.

Halon—Putting the Genie Back in the Bottle[8]

In 1987, Canada was one of the first countries to sign the Montreal Protocol on Substances that Deplete the Ozone Layer. The Protocol, which took effect in 1989, provided a basis for reducing consumption of certain identified ozone-depleting substances, including Halon fire-suppressing agents. In turn, Canada's military forces became one of the first major users of Halon compounds to provide policy and guidelines on the use

---

[8] Ibid., p. 4.

of Halon and to outline the process for the reduction and the eventual elimination of Halon fire-suppressing agents from the fire-protection inventory. In the process of developing policy, the Forces started with the imprecise language of the Montreal Protocol and restructured it into an explicitly worded Halon Management Policy. The major elements of the policy included the following components:

- restriction of the use of Halon fire-suppressing agents to situations where their unique characteristics could not be matched and was required for the protection of resources of critical importance to military operations; and
- a Halon recovery plan that included a procedure of closed circuit removal of Halon from wherever it was in place to protect non-critical resources and to secure the agent in safe storage for use when needed to replace any expended in its newly defined role.

Implementation of the policy required, among other things:

- the immediate de-activation of Halon systems in non-critical applications;
- replacement of critically required systems as soon as replacement/alternative agents became available;
- removal of all Halon portable fire extinguishers (this goal was achieved by 1997);
- provision and maintenance of facilities for the storage of Halon as it was progressively removed from service (the storage bank was built at CFB Borden);
- establishment of criteria for the training and qualification of DND contractors and service technicians who work on fire detection and alarm systems controlling critical use Halon systems;
- monitoring and supporting of research and development of alternative fire-suppression agents and systems;
- participation in the Halon Alternative Performance Evaluation Committee chaired by the CFFM;
- establishment of reporting and investigation procedures and requirements for all Halon releases;
- establishment of an inventory of all Halon holdings; and
- de-activation of all Halon systems in armoured fighting vehicles while not deployed in operations in theatres outside of Canada.

It should be noted that when the Forces started their Halon Management Policy, other agencies, including Environment Canada, had not even begun to come to grips with the problem. As a result, when approximately a year later, Environment Canada struck a committee to develop a code of practice for Halon the DND Halon Management Policy was put on the table. This was the only model available to the committee, and as it proved to be a sound draft policy, much of it was later reflected in Canada's Environmental Code of Practice on Halons.

Halon—A Summary
The use of Halon by the Canadian Forces, which peaked in the late 1980s before being sharply cut back as a result of environmental concerns, provided a high level of protection against fire. Even today, Halon 1301 systems are protecting our serving members on operations around the world, as well as installations and equipment vital to the successful conduct of military operations. The great early promise of Halon 1301 in fixed-pipe systems and Halon 1211 in hand-portable extinguishers, exploited with gusto by the Canadian Forces, was followed by troublesome environmental concerns that, in turn, caused a strong reaction which resulted in them becoming much maligned fire-suppression agents.

In a few instances the navy has installed fine-spray water systems and the army has used FM-200[9] as the extinguishing medium in the engine compartment of some vehicles. Despite all this, Halon has played and continues to play a significant role in protecting the Canadian Forces' operational capabilities. These capabilities were aptly demonstrated by an incident involving two US combat vehicles during the Gulf War. Mistaken for the enemy, the vehicles took direct hits from their own aircraft. The on-board Halon systems immediately activated and saved the lives of eight crew members, four on each vehicle. The genie suppressed the fire and saved the crews, thus demonstrating that, despite its environmental shortcomings, it could perform its magic when called upon.

FIRE-SAFETY EDUCATION

Evidence of the Need
Much of the success of the Fire Service in preventing fires may be attributed, in part, to the fire-safety education programs provided to persons living and/or working on any given armed forces installation. With the cessation of hostilities of World War II and the subsequent reduction of firefighters and fire inspectors, these programs, along with other important functions, were curtailed, which resulted in a significant increase in fire losses for all three services. When we add to this the poor use of building materials, the lack of sprinkler systems in building design, coupled with the use of highly combustible materials, this scenario was a recipe for disaster. Nevertheless, this served to reinforce the need for an all-encompassing fire-safety education program.

Program Details
Defining a detailed fire-safety education program was given top priority. Initially, fire-safety education encompassed a wide spectrum of activities given to diverse groups, such as pilots, ships crews, combat arms personnel, school children, residents of married quarters, boy scouts / girl guides, and the military and civilian work force. Fire-safety education has traditionally included topics such as escape planning, the need for residential smoke alarms, the hazards inherent in cooking with deep fat/oil, fire extinguishers, first aid for burns, juvenile fire setter programs, and the hazards relating to the use of tobacco. A stand-alone program has involved instructing fire wardens who are responsible for fire-safety practices in the building or section where they work.

Practice has shown that whether a fire-safety education program is large or small, it requires extensive year-long planning and must be targeted, continual and measurable. By having measurable items, real problems such as cooking fires or smoking materials are targeted. Fire-safety education topics were aimed at correcting or reducing specific problem areas.

How fire-safety education programs have been conducted has depended to a considerable extent on staff availability. In some fire departments, this responsibility has often been assigned to fire crews as part of their regular duties. In other cases, depending upon the circumstances, fire chiefs have assigned the function to personnel from the Fire Prevention Bureau. Frequently, the program has been bolstered by enlisting others such as community volunteers, teachers, nurses and tradesmen to provide assistance in specific areas. In support of these programs, firefighters have been sent to CFFA to take the fire inspector's course as part of their professional training in which fire-safety education topics have been included.

Since the mid 1970s, NFPA has spearheaded a number of projects in public fire safety for people of all ages, for example, the "Learn Not to Burn Program" for

---

9 See Appendix B for a description.

children from kindergarten to Grade 8. Other notable initiatives have included exit drills in children's schools as required[10] and Exit Drills In The Home (EDITH), a program to help families develop a home escape plan and to practise it.

ENFORCEMENT

Observing and Acting

As a means of providing a measure of surety that the provisions of Fire Safety Codes, standards and regulations, are not overlooked or otherwise abrogated, buildings and facilities are repeatedly visited by fire inspectors and other Fire Service authorities for the purpose of observing conditions of occupancy to:

- identify potential fire and life-safety-related hazards;
- evaluate and research any matter of concern;
- initiate corrective action wherever necessary; and
- bring the importance of practising fire safety to the attention of all concerned.

Fire Investigation

Determining the cause of each fire is a fundamental cornerstone of a fire prevention program. It is of utmost importance in the process of preventing fires from recurring from the same cause. It has been standard practice for all fires to be investigated and for corrective measures to be taken where appropriate.

---

[10] The National Fire Code of Canada and NFPA 101, Life Safety Code.

CHAPTER VIII

MATERIEL

PERSONAL PROTECTION

Turn-Out Gear
Firefighter protective clothing in the early 1940s had remained relatively unchanged since the end of the First World War. Helmets were constructed of Bakelite[1] or leather, complete with the standard elongated brim to afford neck protection or, when reversed, provide protection for the face from radiant heat. The coat was a black, three-quarter length, water-resistant garment named the Petch coat.[2] Hip-length rubber boots that came up under the coat and leather mitts over knitted woollen mitts completed the ensemble. This style of firefighter protective gear is now rarely seen in either Canada or the United States.

In the late 1940s, tests of duck-type fabric for possible use in protective clothing yielded encouraging results. This led to the development of a new style of protective gear incorporating a jacket and separate trousers, which became known as a bunker suit. It was an innovative design that met instant approval from those who tested it. Unfortunately, the bunker suit, including new insulated rubber boots and new helmets which formed part of the new ensemble, was not placed on general issue until January 1960.

During the early 1950s, an asbestos proximity suit was available for ARFF operations. The suit, however, was extremely bulky and clumsy to wear. It also readily absorbed water, which tended to add to its already

---

[1] See Appendix B for definition.
[2] Petch was the name of the company that produced the coats.

considerable weight.[3] It was not a popular piece of protective equipment, but it was one that served the firefighter well and was appreciated when fighting vigorous, flammable liquid fires.

Sometime during the mid 1950s, a bunker suit was produced made of aluminized asbestos. Unfortunately, it had a brief and somewhat unfavourable debut, mainly because of a flaw in the thread used to stitch the suit together. The short-woven, heat-resistant thread fibres had little tensile strength, allowing the seams to break apart. To complicate matters, the material did not wear well and required frequent patching. It was eventually withdrawn from service and never reappeared. This was unfortunate in some respects because the concept of using reflective material had merit; however, its shortcomings were too significant to justify continued use.

Smoke-Eaters and Breathing Apparatus
The title "Smoke Eater" was coined sometime in the distant past, to identify firefighters. Sometimes softened to "Smokies," nostalgia has kept it in the jargon of the Fire Service. It was a legitimate title, since firefighters were expected to perform in smoke-filled atmospheres, without complaint. It was a designation worn with more than a little pride by firefighters, perhaps because the image it generated of the firefighter rushing into what may be a life-threatening situation without regard for personal safety held a certain heroic type of appeal. At the time, however, neither firefighters nor medical authorities were aware of or fully understood the adverse effects that smoke inhalation could have on an individual's health over the long term.

Although it is conjecture at this point, the motivation for the term is believed to be principally because firefighting equipment did not include a functional SCBA until the 1950s.

The issue of breathing apparatus for firefighters was discussed during the Fire Protection Officers' Conference at Ottawa, in April 1941. Participants at the meeting included Wing Commander N.H. Robinson as chairman, Mr. W.F. Clairmont, the Dominion Fire Commissioner, and a fire protection officer from each Command Headquarters of the Air Force. During the meeting, Flying Officer McCallum raised the issue of acquiring of suitable respirators for use by the Fire Service. The AFHQ representative, Flight Lieutenant Snarr, replied that an investigation was being conducted into the possibility of adapting the service respirator to make it effective against carbon monoxide. Following further discussion, attendees unanimously recommended that the breathing apparatus, known as the All Service Mask, a type approved by the United States Bureau of Mine Safety, should be purchased. This decision marked the introduction of an approved breathing apparatus into the Air Force Fire Service. The recommended scale was an assignment of two units per truck, with sufficient reserve canisters to conduct protracted firefighting operations.[4]

The All-Service Apparatus operated on a single canister filled with activated charcoal and a chemical with the trade name Hopcolite. Air was drawn into the canister by the wearer and the chemical contents converted the dangerous carbon monoxide (CO) into harmless CO_2.

There was, however, a serious operational flaw with this unit as there had to be sufficient oxygen in the atmosphere to support life. This shortcoming would prove to be its downfall and it was eventually replaced by the Chemox breathing apparatus.

---

[3] Captain George Cowan (1952–80), *History of the Canadian Military Fire Service* (unpublished, 1990).
[4] Minutes of the Fire Prevention Officers' Conference, April 1941.

The Chemox Self-Contained Breathing Apparatus
Manufactured by Mine Safety Appliances (MSA), the Chemox[5] SCBA became the next generation of breathing apparatus. This self-contained, closed-circuit re-breather unit proved much more effective than the All-Service type. The wearer could rely on the apparatus to generate breathable air, even in the most hostile conditions. The Chemox was the first truly self-contained breathing apparatus used by the Fire Service of Canada's military. Future modifications to this apparatus generally centred on extending the useful working time of the canisters and on the ability of the wearer to start the canisters quickly in low ambient temperatures. Typically, this involved firing one of two of the quick-start candles which would "jump start" the chemical reaction in the canister that was needed to produce the oxygen content needed for breathable air. The candles were also used as an emergency air supply, contributing a quantity of breathable oxygen-enriched air that would provide the wearer with up to four minutes per candle of extra operating time.

Over the years, the Chemox breathing apparatus has proven to be a reliable piece of life safety equipment. Nonetheless, it has had to survive more than its share of controversy. For example, in the early 1960s, there was an Unsatisfactory Condition Report (UCR) raised at the Fire School concerning the reliability of the Chemox SCBA. This UCR action was triggered because of an accident involving a student. The student had started to climb a 40 ft / 12 m ladder wearing a Chemox SCBA, when he somehow ran out of breathable air and fell to the ground unconscious. Fortunately, he sustained only minor injuries, but the ensuing alarm over the incident prompted the UCR action. This action effectively postponed further breathing apparatus evolutions in the fire-training tower for several weeks, causing some firefighter graduates to return at a later date to complete their training.

Despite this incident and the ongoing difficulties many firefighters experienced using the Chemox SCBA, it continued to endure as an effective piece of life-safety equipment. Essentially, the difficulties experienced by some users was caused due to resistance of drawing air through the canister of chemicals when inhaling, with little, if any, assistance from atmospheric pressure due to the closed-circuit design. This was particularly troublesome for firefighters who were in a poor state of physical fitness.

Individual concerns persisted for many years over their reliability. These concerns, as a rule, were motivated either by a lack of training or an individual's misunderstanding of the equipment's operating principles and limitations. This notwithstanding, the Chemox SCBA, with very little modification, remains in service for specific purposes, such as on navy ships where the compactness of the oxygen-generating canisters, relative to cylinders of breathing air, allows for more efficient use of limited storage space.

Compressed Air Debuts
In the late 1960s and early 1970s, the Chemox mask was replaced with SCBA using a cylinder of compressed air with an in-use duration time of 30 minutes. These were supplied by two companies, MSA and SCOTT. The SCBA first of this type were the demand-type[6] models, which were replaced by pressure-demand[7] models to meet updated NFPA standards.[8] Eventually, in order to avoid maintenance problems and to address safety concerns relating to the possibility of mixing parts inherent with the use of more than one

---

5 The name "Chemox" was derived from its operating principle, which involved generating oxygen from chemicals contained in a canister when activated by the moisture in the wearer's breath.
6 See Appendix B for definition.
7 Ibid., p. 4-B.
8 NFPA 1404, Standard for a Fire Department Self-Contained Breathing Apparatus Program.

type, MSA was selected as sole supplier. This was the standard SCBA that, with the benefit of a number of modifications, remained in service until 2000.

The next generation of SCBA was designed to use high-pressure air, which gave the Fire Service a long-waited option to use cylinders of air with either 30 or 60 minutes of in-use duration time. Staying with MSA, all fire halls are now equipped with MSA Model 4500 SCBA, which has enhanced firefighting operations. The 30-minute cylinders are smaller and the use of Kevlar[9] and carbon fibres has made the SCBA unit itself much smaller and 5 to 7 lb / 2 to 3 kg lighter, giving the firefighter more endurance. As well, the cylinders are interchangeable with no added adjustments on the fire ground.

Regulators are now found on the face piece rather than the belt, and improvements in visibility has helped the firefighter perform searches better and has reduced injuries. New safety features have also been added, including quick-fill connections, universal air connections, lumbar support, data logging and data downloading, and a Personal Alert Safety System (PASS). All of the new MSA Model 4500 SCBA used by DND will provide the SCBA user with pertinent data regarding the air-consumption rate, ambient temperature, and alarm activation. Also, all of the above information can be downloaded into a computer for future evaluation. The cost of these units was approximately sevenfold the cost of the first SCBA.

The NOMEX Work Dress
The NOMEX one-piece work dress was introduced in 1970. NOMEX was the trade name for Dupont's new fire-resistive material from which the garments were produced. Resembling an aircrew coverall, the suit marked a new consciousness of firefighter protection, and DND was again at the forefront in this field. This invaluable piece of personal protective gear has evolved to the point where it is worn with pride by firefighters virtually unrestricted, whether on duty or off. This was not always the case, however, and it took some convincing to get authorization to wear the suit outside the fire hall.

Breathable Air—Filling Station
In 2000, with SCBA using air-cylinders of 30 and 60 minute in-use duration time, support in the form of a filling station to refill cylinders after use was needed. To meet this requirement the Fire Service had to upgrade their 3,000 psi / 204 bars low-pressure air supply and storage systems. It was decided to select the MSA Company as the source of SCBA and to upgrade to their new 4,500 psi / 306 bars high-pressure models. One of these units was purchased for each fire hall. The two main components of the units are a compressor and storage cylinders. The compressors were of two types, a Jordair or Mako, each providing 5,000 psi / 340 bars high-pressure air with a pumping rate of 21 ft^3 /min / 0.6 m^3 /min. This provided a safe way of filling air cylinders by providing protection for the user from air cylinder rupture. Each fill station can fill up to three cylinders simultaneously, which reduces the turnaround for returning air cylinders to the fire-ground.

A four-vessel/cylinder system meeting the standards of the American Society of Mechanical Engineers (ASME) was added to provide a reserve of air for filling SCBA cylinders if there was a loss of power or mechanical breakdown of the compressor. Each of the four cylinders holds 520 ft^3 / 14.7 m^3 of breathing air in storage at 5000 psi / 340 bars.

Air Carts
In 2003 air carts were introduced at all fire halls. The air cart is designed to provide four airlines up to 300 ft

---

[9] See Appendix B for definition.

/ 90 m each, and when not supplying breathable air, it can support up to four pneumatic rescue tools with each operating at its own selected pressure. The air cart can use air cylinders rated at either 30 or 60 minutes duration time when used on SCBA. The carts can use the 2,216 psi / 150 bars and the 4,500 psi / 306 bars air cylinders, and come complete with a number of audio alarms to facilitate the safe provision of a continuous supply of breathable air. The air cart is lightweight, compact and readily mobile.

Portable Air Supply System
Due to changes to the NFPA Standard 1500 and the requirement to support firefighter safety, a portable air supply was purchased for each fire department. An MSA design with a universal air-connection, this safety equipment can use either low or high-pressure cylinders of air that have a 30- or 60-minute duration time when used on SCBA. This equipment provides emergency breathing air to firefighters, as well as to victims who are trapped or unable to be removed from a hazardous atmosphere.

RESCUE and EXTRICATION EQUIPMENT

Overview
The rescue equipment used today by firefighters, to extricate persons trapped in a damaged aircraft, a highway vehicle, or as a result of an industrial or other type of accident, clearly reflects the technological advances in the field.

Fifty years ago rescue equipment consisted of a pick axe, a sledge hammer, a crow bar and maybe a jack. Many will remember a huge hacksaw, somewhat like a buck saw, capable of cutting through metal that was carried on the G13 crash truck. Later that saw was updated to a gas-powered Homelite[10] saw. Also carried on the G13 was a crash axe, which looked like a hatchet and had a fully insulated, rubber-coated handle to protect the user if cutting a live, high-voltage wire. It was used to gain access to the cockpit, if the canopy would not open, or to cut through the skin of an aircraft, if necessary. Others will remember responding to automobile accidents, before we had the powered hydraulic rescue tool commonly referred to by the trade name of Jaws of Life, when the only option was to use pry-bars and other like hand tools. However all that has changed for the better. The arsenal of equipment on vehicles today, includes the most advanced equipment that is available.

Air-Bag Lifting System
A requirement for a heavy-lift capability for auto extraction and building collapse was recognized in the mid 1980s and the CFFM decided to invest in an air-bag lifting system for each fire department. These original low-pressure air bags were made by Vetter of Germany and came in 3-, 8-, and 9-ton lift capabilities. Most fire departments found the air bags to be bulky and due to the lack of storage space on most vehicles, they were seldom used. In 2002, a new set of high-pressure-type air bags that are smaller in size, much easier to handle and require less storage space on vehicles, was purchased for each fire department. The high-pressure (120 psi / 8 bars) air bags came in three sizes: 3-, 20- and 35-ton lifting capability.

Air Chisel
The air chisel has been employed with the Fire Service for a number of years. This is used mostly for auto extraction and for aircraft rescue. For many years a 2,216 psi / 150 bars SCBA air cylinder was used with this equipment. Recently a change was made that permits the use of a 4,500 psi / 306 bars SCBA air cylinders or an air cart.

---

[10] See Appendix B for definition.

Hurst Jaws of Life

In late 1977, the CFFM authorized the purchase of six Jaws of Life from Wajax Industries Limited for $6,800 each. In 1978, 31 sets were authorized, with the Code 4 Fire Rescue Company providing servicing. The addition of this first powered hydraulic tool was a great leap in technology and was ten times faster than other hand hydraulic tools. The first sets were made up of a 5-hp motor and a spreader of 18,000 lbs / 8000 kg of force; a cutter was added two years later. In 2001, after many years of service, it was found that most of the equipment required upgrading. New dual-purpose motors were added and several new tools were also obtained, such as rams and a combination tool called Maverick. Another upgrade was made which replaced the heavy spreaders with three lighter spreaders with the trade name of Multi-Lite. With vehicular accidents, there is often a need to extract trapped persons quickly. The Jaws of Life enhanced the ability to achieve this important objective. DND was among the very first to have this tool in Canada.

Power Hawk P-16 Rescue System

Introduced to the fire departments in 1997, this 32-lb / 15-kg 12-volt electric-powered rescue equipment allows firefighters to work quickly to extricate persons entrapped, due to auto and aircraft accidents. Designed and built by Curtiss Wright, it is lightweight, can run in oxygen-depleted atmosphere, requires no hydraulics, and uses a compact, high-performance gearbox developed for the demanding flight-control found on F-16 fighter aircraft. This rescue tool is designed to resist rain and sprays of water from hoselines, and under demanding conditions may be fully submerged for a one-time use. The Power Hawk comes with a variety of attachments, such as power blades, spreader arms, cutters, vehicle power hook-up, a floodlight, a saw and a winch. At time of writing, the navy was in the final planning phase leading to its use on board all of its ships as part of its damage control and aircraft rescue equipment.

FIRE SUPPRESSION

Asbestos Fire Blankets

Asbestos fire blankets, a most unusual piece of equipment for suppressing fire, were commonplace for many years. Most fire vehicles carried at least one, and fuel points and mess halls were similarly equipped. The blankets were designed to be thrown over a burning liquid, such as that contained in a deep-fat fryer. Health concerns regarding the asbestos fibres resulted in them being taken out of service sometime during the 1970s.

Water Fog

In the fall of 1941, the Fog Nozzle Company of Canada Limited introduced an idea that enhanced the method of applying water on a fire by breaking the water into a fine spray, commonly referred to as water fog or just fog. These fog nozzles were referred to as Navy B head nozzles; they were initially designed to combat fires in ships' boiler rooms. Trials conducted at RCAF Station Uplands[11] proved that low pressure water-fog devices attached to a suitable applicator could be effective in combatting flammable liquid fires. The success of these tests resulted in Griswold fog nozzles being placed on every crash tender.

A widely used term for these nozzles was fog applicators. These applicators were essentially a pipe shaped much like a hockey stick. The longer section of the pipe had a coupling for attaching to a hose with an attendant shut-off valve, while after the curve, the shorter section was fitted with a discharge head in the form of a half-sphere with many small holes drilled at opposing angles to cause each high-velocity stream to collide, one with the other, to fracture the streams into a fog. The main purpose of the curve was to cause the

---

11   Appendix to the minutes of the Fire Prevention Officers meeting held in Ottawa, May 25–28, 1942.

discharge head to be on the same plane as the longer section of pipe, in order to provide for even application of fog over the surface of a liquid-fuel fire. Initially they were manufactured in 5 ft / 1.5 m lengths and 10 ft / 3 m lengths. In the early 1950s, the Air Force also used a 3 ft / 1 m length and added a "D" handle just in front of the coupling to enhance mobility in close quarters.

With foam not yet widely available, some fire departments practised and employed special tactics for entering liquid-fuel fires using the fog applicators where life was at risk.[12] Firefighters were trained to place their left hand around the pipe just behind the discharge head and to grasp the "D" handle with their right hand. With the helmet on back to front, the discharge head was brought up towards the face until some of the spray struck the underside of the brim of the helmet and thence onto the face. With two applicator men and one rescue man, these teams were able to work while totally surrounded by fire. It was, however, very difficult to protect their backs, which frequently resulted in the lower portion of the Petch coats being burned off. It was a dangerous practice that would not be countenanced today.

The advantages of using fog instead of straight streams of water pushed the development of fog- producing equipment that would be more versatile than the fog applicators. This eventually resulted in a nozzle called a mystery nozzle.[13] This permitted the use of indirect firefighting tactics in certain circumstances to suppress fire in buildings, as espoused by Lloyd Layman, the great firefighting tactician and equipment innovator, who was coincidentally conducting similar trails at a US naval dockyard installation in Maryland. He became an advocate of water fog for firefighting and its special application in the indirect method of structural firefighting. The results of his trials were published in two notable books[14] that became the firefighter's bible for structural firefighting, including size-up methods and tactical extinguishment manoeuvres.

Firefighting Foams
Foam achieved dominance as a fire-extinguishing agent rather quickly after its introduction a half century ago and still remains the primary weapon in the firefighter's arsenal for fighting flammable liquid fires. The early 1960s marked the advancement and improvement of foaming agents and their effectiveness to extinguish and provide a vapour seal in all manner of flammable liquid fires. For many years, there was not an effective extinguishing agent for some materials, such as liquids containing a significant percentage of alcohol; however, since the early 1980s, alcohol-resistant foam[15] has been commercially available. Foam products have also come into limited use in structural firefighting and in fighting wildfires.[16]

General Characteristics of Foam
Foam is essentially an aggregation of small bubbles used to form an air-excluding, vapour-suppressing blanket over the surface of a fuel. All types of foam concentrates now in use are in the form of a liquid, which is proportioned with water for generating foam bubbles that, in their multitudinous numbers, constitute a foam blanket. Firefighting foams are usually defined by their expansion ratio, which is the ratio of the final volume of foam produced after aeration, compared to the original volume of the water/foam solution before

---

12 Memoirs of Flying Officer Doug Stevenson (1946–63).
13 See Appendix B for definition.
14 Layman applied military tactics to the fire service when he was a fire chief in West Virginia and initiated pre-fire planning and published *Fundamentals of Firefighting Tactics* in 1940. It is still the basis for many of today's procedures. He established and led the Coast Guard Firefighting School in Baltimore and was Director of the Fire Office, Federal Civil Defense Administration. His second book is titled *Attacking and Extinguishing Interior Fires*. It was published in 1955 by the NFPA.
15 See Appendix B for definition.
16 Ibid., p. B-5.

adding air. They are arbitrarily subdivided into three ranges of foam expansion ratios: low-expansion foam, expansion up to 20:1; medium-expansion foam, expansion 20:1 to 200:1; and high-expansion foam, expansion, 200:1 to1000:1.[17]

Some firefighting foam is generated from solutions with very low surface tension which enhances the penetration characteristics of the mixture. Foam of this type is useful where Class A combustible materials are present. In such instances, the water solution draining from the foam cools and wets the solid combustible.

The most common foaming agents are: Protein Foam (PF); Aqueous Film-Forming Foam (AFFF), Fluoroprotein Foam (FPF), Film-Forming Fluoroprotein Foam (FFFPF), and Alcohol Resistant Foam (ARF). AFFF and MXF concentrates have been the two most widely used by the Canadian Forces since the early 1970s.

Establishing and Maintaining a Foam Blanket

Foam breaks down and vaporizes its water content under attack by heat and flame. It, therefore, must be applied to a burning liquid surface in sufficient volume and rate to compensate for this loss, with an additional amount applied to guarantee a residual foam layer over the extinguished liquid. Foam is unstable and may be broken down easily by a physical or mechanical force, such as a water-hose stream. Certain chemical vapours or fluids may also destroy foam quickly. When certain other extinguishing agents are used in conjunction with foam, severe breakdown of the foam may occur. Turbulent air or violently uprising combustion gases from fires may divert foam from the burning area.

Protein Base Foam Concentrate

PF began to be used in the early 1950s and remained in use into the 1970s, during which time it was the primary foam used to combat aviation fires and, in the broader sense, all fires involving liquid fuels. PF concentrate was inducted into the water stream using an in-line inductor calibrated to introduce foam concentrate at a rate that resulted in a liquid mix that was 6% foam concentrate and 94% water. An air-aspirating nozzle was required to introduce sufficient air into the mixture to produce foam with an expansion ratio adequate to blanket a large area. Beginning in the late 1950s, air blowers were used on crash vehicles to inject air into the water/foam concentrate mix to improve the quality and expansion ratio of the foam.

PF concentrates produce dense, viscous foams of high stability with good resistance to "burn-back."[18] They are non-toxic and biodegradable, a fact not lost on firefighters who discovered the excellent fertilizing qualities of this product when cautiously spread on lawn or garden!

Dry Chemical

The introduction of sodium bicarbonate dry chemical (DC) about 1950 was seen as a plus in fighting flammable-liquid fires. In fact, combining the potential of PF and fire extinguishing agents should have proved to be a firefighter's dream; however, it initially caused exasperating problems. Specifically, when it was used in conjunction with protein foam, the sodium-based powder broke down the foam blanket and eventually the foam's vapour sealing qualities. This was due to the DC being treated with magnesium base material, which is a soap-like substance that destroyed the foam bubbles. This reaction effectively curtailed the combined use of these two agents for a time. To alleviate this problem, the DC was treated with silicone, rather than the magnesium base material. The purpose of both treatments was to protect the DC

---

[17] NFPA Handbook.
[18] See Appendix B for definition.

particles from absorbing moisture. The new product was designated "foam-compatible" DC, and while it did reduce the deleterious effect of the dry-chemical on the foam bubbles, it did not completely eliminate the problem. Work conducted by the Ansul Company, working in conjunction with the US Navy, developed a more complete solution to the problem in the early 1960s, by using potassium bicarbonate (PK), instead of a sodium-base powder. The US Naval Research Laboratories (NRL) named the product Purple K, which became the common term used.

AFFF

Product development on AFFF was initiated in 1962 by the 3M Company following a meeting with the NRL. The NRL had noticed the film-spreading capability and vapour seal-ability of fluoro-surfactants on petroleum fuels. The first commercial product, identified as FX-1831, was introduced in 1963 after testing with the Navy. Success, however, was limited, due to design considerations that required a 25% induction rate. The first successful product that met customer requirements was identified as FC-1941, which 3M introduced in 1965. This product was effective at a more practical 6% induction rate. In 1966, the 3M Company received approval for the Light Water trade name, by which the product became widely known.[19]

The development of AFFF improved the fire-knockdown and fire-suppression qualities of foam. AFFF solutions had fast-spreading and levelling characteristics and provided a very good surface barrier to halt vaporization. Its rapid knockdown capability pretty much relegated DC to the limited role of extinguishing three-dimensional fires.[20]

Clearly a superior foam agent, AFFF replaced protein foam as the main firefighting agent for hydrocarbon fuels. It came into widespread use in DND coincident with the 1972 Oshkosh 3800 L / 840 G crash truck, which was designed to produce foam using this synthetic foam concentrate which has quicker fire-control capabilities. The arrival of this vehicle would, in time, replace the G19 Alvis and the G19 Sicard and end the use of PF.

MXF

MXF with an expansion ratio of 20:1 to 200:1 was first tested by the military firefighters at CFB Lahr, West Germany, in either late 1969 or early 1970. This product could be used on either class A or class B fires. It was especially effective when introduced into confined spaces using a medium- expansion nozzle which yielded an expansion ratio of approximately 75:1. The trials were successful, and it was decided to carry the agent on the in-service pumpers and to incorporate a foam tank in the new pumpers that were in the planning stage at the time.

The new pumpers had been in service a few years when a fire occurred in a basement storage room of an apartment building used as permanent married quarters (PMQ). The basement was full of smoke; the basement windows were opened to ventilate the area; the seat of the fire was located and MXF was used to fill the room involved. After a few hours, the foam blanket was gone, with little or no water damage, and the origin of the fire was clearly visible.[21]

Back in Canada, in the early to mid 1970s, existing pumpers on some bases were being retrofitted with similar packages. In 1980, the military specifications for pumpers were changed to include an MXF package. In 1982, Bickle-Seagrave was awarded a contract to build 12 pumpers with built-in foam systems.

---

19 The 3M Company Book, *A Chemical History of 3M*, 1933–90.
20 See Appendix B for definition.
21 See Chapter X for greater detail.

Thermal-Imaging Cameras
The Navy Damage Control Teams on board ships first used thermal-imaging cameras (TIC) in the early 1980s, and a small number were purchased for HAZMAT operations. They were eventually purchased for all fire departments in 2002. The TICs now in use are easy to use and have a large screen and heat-seeker indicator. Most fire departments have two TICs to support ARFF and structural fire-suppression operations. This equipment has already been credited with saving lives and reducing property damage.

Summary
Over the years, the many marked improvements to firefighter safety and operational effectiveness, quite apart from the mental and physical capabilities of its members, are closely interrelated to advancements in materials and equipment. Consider:

- In 1941, perhaps the first significant step forward occurred when the ability to produce fog was developed, thereby adding a new dimension to water as a fire suppressant nearly a decade before foam became available.
- In 1941, the All-Service Mask was introduced in limited numbers, although the risky practices that gave rise to the "Smoke Eaters" synonym for firefighters continued for many years.
- In 1950, foam and DC began a long run as primary extinguishing agents.
- In 1950, the Chemox SCBA, which allowed firefighters to work in hostile, oxygen-deficient atmospheres, came into service.
- In 1960, the long, black Petch coat, hip-length rubber boots, and Bakelite helmets gave way to the jacket and trousers style bunker-suit and knee-length rubber boots and helmets, providing improved protection and mobility.
- In the late 1960s and early 1970s, the Chemox mask was replaced with SCBA, using a cylinder of compressed air to enhance firefighter performance.
- In 1970, the NOMEX one-piece work dress, tailored from fire-resistive material, was introduced, upgrading firefighter safety and allowing for quick transition from a normal state of stand-by-readiness to an emergency-response posture.
- In 1977, the Jaws of Life hydraulic power rescue tool was a quantum leap forward from a mix of hand tools that had been the best available for many years. It allowed a tenfold reduction in the time required carry out some critical rescue operations.
- In the mid 1980s, an Air-Bag Lifting System was introduced for use in certain rescue operations.
- In 1997, the Power Hawk P-16 Rescue System was introduced which enhanced firefighter's ability to quickly extricate persons entrapped due to auto and aircraft accidents.

Picture Gallery VIII

Introduction

As things began to unfold with this photo gallery, it became increasingly evident that photographic records in this area did not extend backwards in time beyond the last decade or two. An exception to this was evident in a photograph of a Fire Prevention Week display that had been set up on the apparatus floor of the No.11 Supply Depot Fire Hall in Calgary, about a half century ago. It shows a rather extensive outline of what was front-line equipment at that time. The caption to the photo identifies and describes as many of the items as possible.

The venerable hose-reel cart has served for many years, virtually since the beginning of the Fire Service.

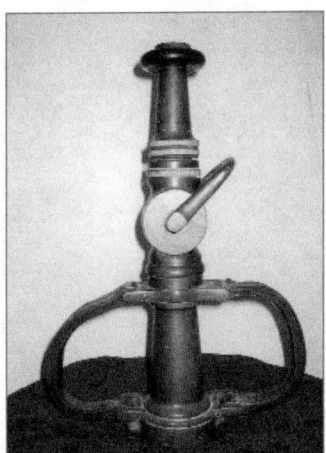

The straight-stream nozzle. Of solid brass construction with a 1-in/2.5-cm tip, it weighed close to 20 lb / 9 kg. Very heavy, but virtually indestructible.

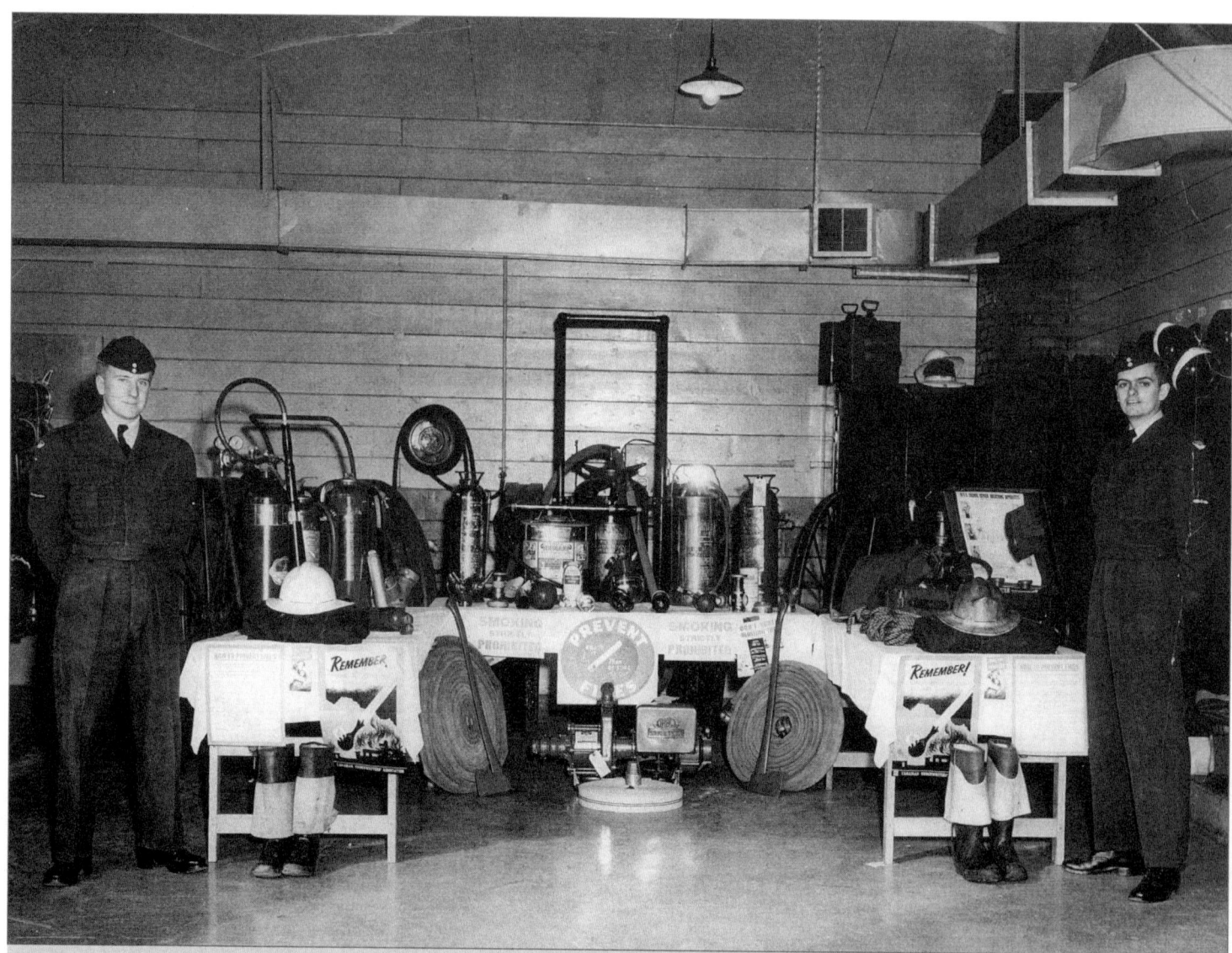

Fire Prevention Week Equipment display of a half century ago at RCAF No. 11 Supply Depot, Calgary. LAC Lee Allen and LAC Lorne MacLean await the arrival of a group of Guides and Scouts.

Perimeter
A 300-lb/135-kg DC wheeled-unit extinguisher. The pressure reduction valve, pressure gages, curved handle and broad faced wheels may be seen.

A 40-G/190-L chemical-foam wheeled-unit extinguisher. The tall handle, the tank with operating instructions on its faceplate, the screw cap handles, a slender stem extending through the cap with a hand control for controlling the cap on the inner chamber, and the broad-faced wheels may be seen.

This extinguisher had two chambers, each containing a different liquid chemical mixture. To operate the extinguisher, the cap on the inner chamber was opened and the extinguisher tipped down to cause the contents of this chamber to spill into the outer chamber. Mixing the two chemicals blends caused a chemical reaction which, in turn, caused a build-up of pressure and yielded a good quality foam.

Two pump-tank extinguishers and two disused helmets atop a metal locker.

A hose-reel cart with a load of 2½-in/65-mm cotton, rubber-lined hose and a straight-stream nozzle. Other equipment, such as a hydrant key, hose spanners and hose straps, were carried in a tool box mounted forward of the reel.

Table
A 15-lb/7-kg CO_2 extinguisher.

A 20-lb/9-kg DC extinguisher.

A 30-lb/14-kg DC extinguisher

A 4-lb/2-kg DC extinguisher, lying flat on the table. This type did not have a hose and nozzle, just a spigot, which may be seen at the bottom.

A CO_2 pressure cartridge for a 30-lb/14-kg DC extinguisher.

A 2½-in x 2½-in x 2½-in / 65-mm x 65-mm x 65-mm Siamese connection.

A 1½-in x 1½-in x 1½-in / 38-mm x 38-mm x 38-mm wye connection.

A 2½-G/11-L chemical foam extinguisher. Its design was similar to the wheeled-unit, except that the inner chamber had a lead stopper held in place by gravity. To operate the extinguisher, it was inverted to the upside-down position to allow the contents of both chambers to mix.

A 2½-in/65-mm mystery nozzle.

A 5-lb/2.3 kg-CO_2 extinguisher (round bottom and horn showing).

A 5-G/28-L backpack-style water pump-tank extinguisher for natural-cover firefighting. The pipe-like slender nozzle was used to generate pressure as a pump for a bicycle tire, except that the pressure stroke was as the outer shell was pulled towards the operator, rather than the push-away stroke of a bicycle pump.

...cont'd on page 144

...cont'd from page 143

A cartridge of water-softening agent, in glycerine form, for use in the hydroblenders fitted on pumpers.

A tear-drop shaped extinguisher containing CTC under pressure. It was designed to be hung on a ceiling in small shops and to operate automatically when exposed to elevated temperatures.

A small CTC extinguisher, size not determined.

A 1-qt/1-L CTC extinguisher (bottom showing), with a hand-operated pump to expel the agent. Due to its high toxicity, especially when exposed to high temperatures which converted it to phosgene in normal atmospheric pressure, it was deemed unsafe and was withdrawn from use in the late 1970s or early '80s.

A CTC extinguisher, size not determined.

A 5-G/23-L water pump-tank.

A $2^1/_2$-lb/1-kg CO_2 extinguisher.

A $1^1/_2$-in/38-mm mystery nozzle.

A $2^1/_2$-G/11-L loaded-stream extinguisher. It contained a liquid fire-retarding chemical mixture with a cartridge of CO_2 to expel the agent.

A $2^1/_2$-in/65-mm adaptor used to change from one thread specification to another.
A $2^1/_2$-in x 1v-in x $1^1/_2$-in / 65-mm x 38-mm x 38-mm wye connection.

A hose spanner with just the chisel-shaped end showing.

A Chemox SCBA shown in its carrying case with the cover open. The tops of two canisters of chemical can be seen.

A coil of natural-fibre rope. This type of rope was phased out in the late 1970s or early '80s and was replaced with similar sized synthetic rope which had greater strength.

In the foreground at both sides of the table is a folded Petch coat with a helmet on top. The helmets were made of fibreglass. They were both strong and relatively light, but with use, they tended to fray and become ragged as the fibres separated.

Floor
A pair of hip-length rubber boots at each leg of the U-shaped table. When pulled up fully, they extended above the hem of the Petch coat.

A flat-head axe leaning against 50-ft/15-m length of cotton rubber-lined $1^1/_2$-in/38-mm hose on the left side, and an axe pick-head against a 50-ft/15-m length of $2^1/_2$-in/65-mm cotton rubber-lined hose on the right.

In the centre, a 100-ft/30-m length of linen hose with a strainer for the hard-suction hose of a forestry pump.

An IEL forestry pump.

Various hand tools, most of which were introduced in the late 1970s and early '80s.

HAZMAT decontamination kit, introduced in the late 1980s or early '90s.

Confined space kit, introduced in 2003.

High-angle rescue kit, introduced in 1990.

Vetter air-bag lifting system, introduced in 2003.

WAJAX pump, introduced in the late 1960s or early '70s.

CHAPTER 8 • MATERIEL

MSA 30-minute SCBA, introduced in 2000.

MSA 60-minute SCBA, introduced in 2000.

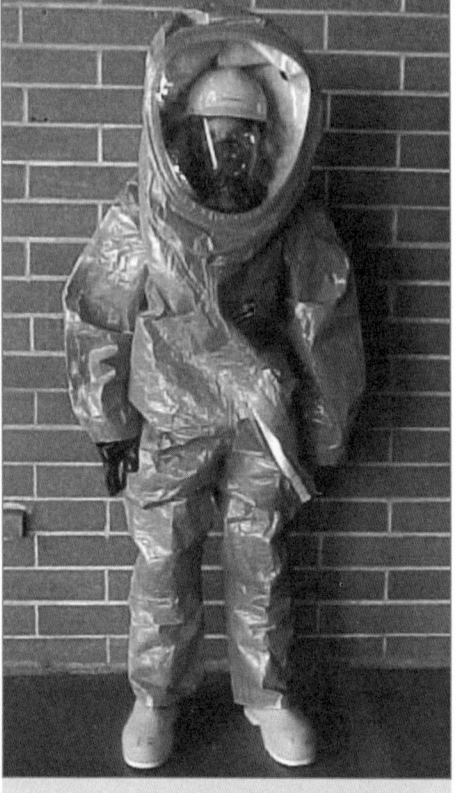

Fully enclosed HAZMAT suit, introduced in 2003.

Breathing-air filling station and Cascade air-storage system, introduced in 2000.

Breathing-air compressor, introduced in 2000.

Breathing-air cart, introduced in 2002.

Power Hawk rescue equipment, introduced in the mid 1990s.

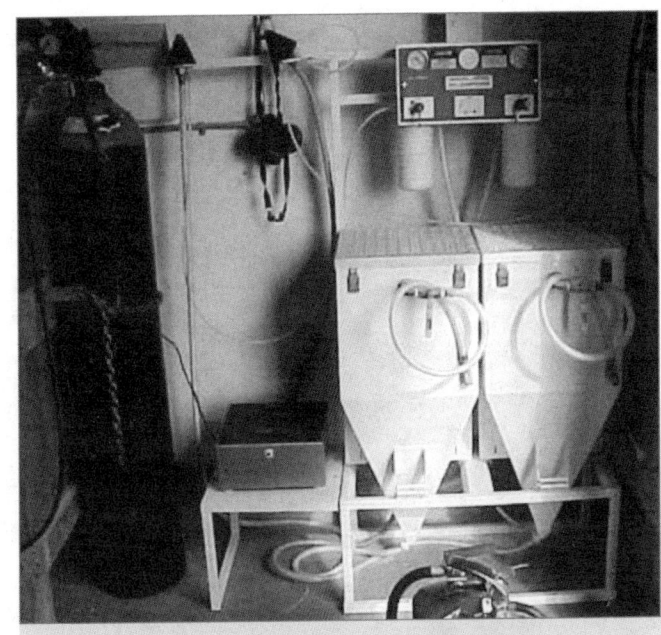
DC recharging machine, introduced during the 1990s.

Hydro-static tester.

Hurst Jaws-of-Life. Introduced initially in the late 1970s, this model came into use in 2000.

Air chisel.

CHAPTER 8 • MATERIEL

Spill kit for HAZMAT.

Air pump for HAZMAT.

Plugs, HAZMAT kit.

Plugs, HAZMAT kit.

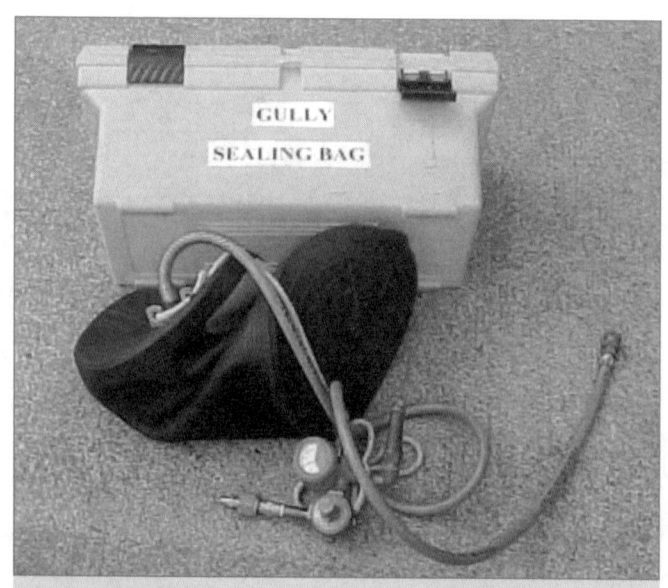
Plugs, HAZMAT kit.

CHAPTER IX

APPARATUS

BACKGROUND

Overview
Our knowledge of mankind's early struggle to control fire is sketchy to say the least. Consequently, some assumptions must be relied upon. In early times, if a fire was not immediately doused with water using some handy container, it would have to be left to burn itself out. Little could be done with the small amounts of water that could be stored in pottery cooking vessels, or other similar receptacles that were in general use, providing they happened to be filled with water when a fire started. Buckets had yet to be developed.

Romans and Greeks Lay the Groundwork[1]
It has been recorded that the Romans guarded against fire by ordering every householder to keep buckets, syringes, hooks and mops ready. These types of household tools were the chief means of fighting fires for many years, perhaps centuries, before the development of pumps and fire engines. Long-handled hooks were used to pull burning roofs to the ground where the fire could be put out more easily, much like the tactics of using pike poles today. Other hand equipment included large hammers, simple ladders, and the humble bucket.

Early inventors, however, were busy developing new ideas. During the Roman occupation of what is now England, a man called Pliny the Elder invented a simple siphon-squirt or syringe that sucked up water from a container and squirted it at the fire. This device,

---

[1] Arthur Ingram, *A History of Firefighting and Equipment* (Chartwell Books Inc., Secaucus, New Jersey: 1978).

dubbed a sipho, was used in Britain with few changes up to the 17th century. Later on, a Greek inventor built a brass pump with an air chamber and two pistons working alternately to produce a continuous stream of water. Later inventors added a discharge pipe, which was fitted with a swivel joint, so water could be directed at the fire. This device became known as the gooseneck, which is still in use today.

Following the Footsteps of the Romans
The development of a military fire service, along with the required equipment, was somewhat akin to the Roman situation. Strict rules and regulations along with basic training for volunteers, backed up with a minimum of equipment, was the common practice. In the early years, equipment was primarily hand-operated extinguishers containing water. Buckets of sand were placed in the same general location as the extinguishers. On some military stations where hydrants were available, hose carts equipped with hose, nozzles and spanners, pulled by fire-piquets, were the last line of defence, if hand-held extinguishers, operated by the building's occupants, failed to put out the fire.

In Canada, the first fire truck belonging to the Air Force was acquired from the RAF in about 1920 and was located at Camp Borden. The second procurement of trucks was two pumpers in 1939. These trucks represented the complete mobile fire protection for the Air Force, and were located at Camp Borden and Trenton. All three trucks were small by today's standards, but were more or less state of the art at the time. The 1939 trucks were equipped with suction hoses, hose reels and ladders, and at least one portable fire extinguisher.

EVOLUTION IN MODERN TIMES

The Immense Impact of World War II
The declaration of war in the fall of 1939 soon had a major impact on the purchase of equipment of many kinds and for many purposes, including that required by the Fire Service. Bases were being built all across the country to support the rapid build-up of the Navy, Army and Air Force. The development of infrastructure was especially rapid in response to the need to train aircraft pilots from Canada and other Commonwealth countries. All of which triggered a huge demand for more fire protection.

The Air Force alone purchased more than 100 pumpers and more than 250 crash trucks. Initially the pumpers were identified as Code 30 and the crash trucks as Code 33. These vehicles were later re-designated as G10 and G15, respectively, and remained in service until the early 1950s. In 1941, the Air Force purchased a ladder truck, which was stationed at Gander; however, very little is known about the truck or its fate.

The experience gained from operating ARFFV, which relied on water or water plus foam concentrates, in minus 30F°/34C° temperatures along the NSR and in many parts of Canada prompted the Fire Service to look at alternative firefighting agents.

Early Development—Aircraft Rescue and Firefighting Vehicles
From the beginning of flying operations there was an obvious need for a crash response vehicle with an efficient off-road capability. Crash-site conditions were often hostile to the passing of heavily laden motorized vehicles. The first attempt at manufacturing an off-road vehicle began in 1937 by specialists at Camp Borden who were determined to resolve this problem; they assembled a half-track crash tender on a Ford chassis complete with a firefighting package. It was a good idea; however, the combined effect of lack of vehicular speed and a shortage of development funds resulted in the project being abandoned before it became a workable production model. Many subsequent attempts

to produce an all-terrain vehicle were unsuccessful as each fell victim to the fact that fire-extinguishing agents are extremely heavy and would either sink or slow the machines when they were forced to leave firm ground. Nevertheless, this 1937 project was an imaginative effort that faired no worse than many later attempts to manufacture a similar off-road vehicle. The experiment, however, did have something of a positive spin-off, as the chassis provided the foundation for the venerable Code 33/G15 Crash Tender.

Regardless of the progress in vehicle design, there were still equipment shortages and some units still waited a considerable time before taking delivery. Typical of the times, the Opening Deficiency Report on No. 10 Air Observer School in Chatham, New Brunswick, signed by Flight Lieutenant P.S. Snarr, Fire Prevention Officer at AFHQ,[2] directed that Motor Transport vehicles be provided with fire equipment to serve as a crash tender until one was made available. This situation was repeated in many other locations until vehicle production caught up with demand.

Experimenting with Dry Chemical
In the late 1940s, DC, the heralded panacea of extinguishing agents, had only recently been developed. Initial use of a DC extinguishing agent was confined to fire extinguishers that included first-aid, hand-held units and 300 lb / 135 kg wheeled units. However, innovative firefighters immediately began experimenting with the fire-suppression potential of DC when married to a suitable fire truck.

The AFFM's staff carried out the trials program at RCAF Station Uplands. The experimenting began and involved mounting a 300 lb / 135 kg DC unit on a truck and using it to combat a liquid fuel fire. During the trials the concept showed promise and plans were made to convert six G15s to G17s, each fitted with four of the 300 lb / 135 kg units for a total of 1200 lb / 545 kg of DC. This project was carried out between 1948 and 1950.

Conducting the trials were Warrant Officers Bill Walker, Fred Sacho and "Bull" MacFadyen. Witnessing the event was Flying Officer Bert Quinn of AFHQ and representatives from the USAF, United Airlines, the United States Navy and the US Fire Underwriters Association. The DC trials were an unprecedented success, particularly when a modified vehicle was able to carry four of the 300 lb / 135 kg units. Despite some inherent weaknesses, DC soon made its way to the front line of aircraft crash and flammable-liquid firefighting. Due to their close involvement with sodium bicarbonate and the surrounding publicity, for a time, the firefighters were referred to by pilots as "soda jerks." Not, however, in a derogatory fashion as Gerald Waring noted in the *Trenton Contact*.[3]

The Impetus for Change
Up to the early 1950s, the fleet of crash fire vehicles more or less matched the potential threat, in that the fuel load carried on most aircraft was small. However, things were changing, aircraft were getting bigger, and passenger and cargo loads were getting bigger. The fire vehicles in operation at the time included the G10 open-cab Pumper; the Class 155 open-cab crash vehicle, the G15, the Class 125, and the G17.

RCAF Station Edmonton / Edmonton Municipal Airport, where the Fire Service provided fire protection services, is a good example of the conditions that prevailed at the time. The aircraft using the airport included the RCAF North Star, the privately owned York, which were Lancaster bombers converted for use as freighters, and the USAF Globe Master, which were

---

2 AFHQ File 926-15-1 May 21, 1941.
3 *Trenton Contact*, 1951, p. 5.

hauling heavy equipment to the far north for construction of the Distant Early Warning (DEW) Line of radar stations. This latter operation was on a 24-7 footing with landing aircraft, refuelling hot, loading cargo, changing crews and taking off. The USAF protocol was to have a fire truck standing by during hot refuelling. The fire department was ill equipped for the job with open-cab trucks that had no cold-weather protection in conditions where the temperatures were frequently in the minus 20°F / minus 29 C° to minus 30°F / minus 34C° range.

A Serious Shortcoming Finally Addressed
Outdated and rapidly heading toward obsolescence, the venerable G15 was still being used as a front-line vehicle in the early 1950s, but was inadequate for the role. Adoption of the NFPA Standard 403[4] by DND gave the AFFM the ammunition he needed to petition for new fire vehicles. This standard contained the minimum requirements for aircraft rescue and firefighting services at airports. It also designated a numbered category for an airfield to indicate the type of fire vehicle, its firefighting capacity, and other equipment needed to maintain the airfield's category. The airport category was based on the largest aircraft scheduled to operate out of the airport, and the RCAF stations were initially given one of three categories: five, six or seven, with seven requiring the highest level of fire protection. Assignment of a numbered category provided the fire marshal with a yardstick for allotting personnel and equipment in an orderly manner.

In 1948, the first post-war northern exercise, code-named Operation Sweet Briar, revealed several inherent weaknesses in the type and quality of equipment available for crash firefighting in the north. The G15 crash tenders main fire-extinguishing agent was water-based, and the absence of insulation allowed the water lines to freeze. An additional problem was the limited capability of the vehicles firefighting package to extinguish flammable liquid fires.[5] Armed with the NFPA Standard 403, the fire marshal was able to obtain funds for a foam-producing vehicle to replace the G15.

Introduction of DC
DC provided the firefighter with a weapon that had outstanding flame knock-down capability compared to the existing foam or water-fog agents. Accordingly, there was an immediate desire to deliver this agent rapidly and in large quantities to the scene of a fire or aircraft crash site. Innovation was the order of the day and, with the help of several construction engineering plumbers, a G15 Crash Tender was modified to carry four 300 lb / 135 kg DC wheeled extinguishers. Although this modification did not prove to be a complete success, it did open the way for the future evolution of a truly efficient DC crash truck.

Firefighting tactics also had to be modified to maximize the potential of the new agents. DC's main weakness was its inability to provide a vapour seal over the surface of a liquid fuel. As a consequence, the vapour emitting from the fuel was prone to re-ignition by flames left flickering in hidden pockets, or by smouldering, solid combustible material. This reaction, often violent, became known as a flash-back. Even after the demanding firefighting techniques unique to this fire-suppression agent were mastered, flash-back fires were frequent.

The First DC Crash Truck
The first venture into a new postwar crash truck was the development of the G18 DC crash truck in 1951. It was a questionable concept and yielded poor results. The concept being pursued was to combine crash and structural capability in the one vehicle. The truck was

---

[4] NFPA 403, Standard for Aircraft Rescue and Fire-Fighting Services.
[5] *Trenton Contact*, 1949, p. 10.

built by Bickle-Seagrave of Woodstock, Ontario, and the Ansul Company of Marinette, Wisconsin. The result of this collaboration was a vehicle equipped with two steel tanks, each with the capacity to contain 2000 lb / 900 kg of DC. It also had eight 400 ft^3 / 11 m^3 cylinders of nitrogen.

The design of the G18 was basically flawed, however, because of the immense weight of its firefighting package. As well, recharging the vehicle after use was both difficult and slow, and the vehicle's handling was very cumbersome and difficult for even the most experienced drivers. As one firefighter remarked, it took a ten-acre field to turn it around—not exactly a driver's dream truck. To its credit, the G18 did have an impressive fire-killing package when it did arrive at the fire. Also on the plus side, technical knowledge and innovative techniques learned while developing the G18 were invaluable. The payoff came when it was time to engineer the next generation of DC trucks when it was acknowledged that installing 4000 lb / 1800 kg of DC on an initial crash response vehicle, along with all the other equipment, was impractical.

Due to the backbreaking work involved in recharge operations and unwieldy handling characteristics, the G18 became known as "the monster." The death knell sounded for the G18 when, while been driven from Ottawa to Camp Borden, it suffered a tire blow-out and crashed into a ditch. It was never returned to service.

INTERNATIONAL STANDARDS

ARFF—Calculating Requirements

Up until the 1970s, the method of determining ARFF requirements was based on rather imprecise reasoning, which was about to undergo a quantum change. The catalyst for this was the committee responsible for NFPA 403 Standard for ARFF Services. This committee had been gathering statistical data from hundreds of crashes all over the world which, in turn, led to the development of a method of calculating requirements based on what the collected data reflected and on extensive controlled testing (see Appendix H). Adoption of this international standard gave the fire marshal a strengthened position in delineating ARFF requirements.

Calculating Response Times

Historical data of airport crashes indicate that 85% occurred within the Rapid Response Area (RRA), which is the runway length plus 500 m / 1600 ft at each end and 150 m / 500 ft outwards from the centreline of the runway. NFPA 403 specifies that response time of the first responding ARFF vehicle to reach any point on the operational runway must be two minutes or less, and to any point remaining in RRA, must be no more than two and one-half minutes.

Satisfying Vehicular Requirements in Principle

The challenge was to combine the right firefighting package with the right vehicle chassis. In order to achieve this aim, it was imperative that the fire marshal be the design authority for the firefighting package. The CFFM staff, working in co-operation with staff of the Vehicle Engineering Directorate, with the full and active support of the latter, achieved this in the late 1970s. In the late 1970s or early 1980s, these two groups faced the issue of foam vehicle downtime from a broader perspective. Clearly, a service life of 18 years for fire vehicles was manageable for pumpers, but was demonstrably too long for the more complex ARFFV, which was further complicated by the extensive use of chemicals that had a deleterious effect on many components. It was with a great feeling of satisfaction that the service life of ARFFV was reduced to 12 years.

Attention was turned to two critical operational problems that needed to be addressed, both associated with excessive downtime involving ARFFV, particularly during the later years of the vehicle's service life when they became extremely difficult to maintain. There were serious spin-off complications, in that when a vehicle was out of service, the fire chief was morally obligated to notify the air-traffic control authorities of the unavoidable reduction in the ARFF Category. The bigger and more complicated crash trucks became, the more maintenance they required, resulting in lower levels of protection for longer periods. The solution was to provide a spare foam vehicle, which was done in the early 1980s.

Minimum Number of ARFF Vehicles
The minimum number of ARFF vehicles provided at each airport for each category of airport, as specified in NFPA 403, is shown below with the proviso that consideration shall be given to the provision of an additional vehicle or vehicles in order that minimum requirements are maintained during periods when a vehicle is out of service.

Figure IX-1
Minimum Number of ARFF Vehicles

Airport Category	1	2	3	4	5	6	7	8	9	10
Minimum Vehicles	1	1	1	1	2	2	3	3	4	4
Spare Vehicle	1	1	1	1	1	1	1	1	1	1
Totals	2	2	2	2	3	3	4	4	5	5

The ROSTER

Introduction
The roster of fire apparatus has been arranged in chronological order, keyed to the date of manufacture. All pertinent information that could be obtained for each piece of apparatus has been included. In some cases this is quite extensive, while in other cases, precious little detail was found. Essentially, the same principle was followed respecting photographs in that the best that could be found have been used, even though they are not all of high quality. The vintage 1918 fire house is inserted as a means of illustrating and capturing the flavour of the times.

1920 The First Canadian Fire Truck
This vehicle was originally owned by the RAF during their tenure at Camp Borden.

The Nashwack Fire Boat
One of three fire boats that were in service in Halifax Harbour during World War II, it was manned by Navy personnel until 1946, after which manning was by DND civilian firefighters.

1938 Ford Crash Tender
This prototype crash tender, with the designation of N-340, was built for the Air Force and underwent testing at Camp Borden. It was intended to be an all-terrain type vehicle; however, for reasons unknown, it never got beyond the testing stage. In the photograph gallery it can be seen that a funnel was mounted on the tailgate. It is assumed that this was used to feed a powdered foam chemical into a mixer to blend it with water to produce foam bubbles with sufficient expansion to place a vapour-sealing blanket over a liquid fuel. The vehicle also carried large units of CO_2 for delivery through hose-lines mounted on reels.

1939 GMC/American-LaFrance Pumper
This was the first pumper at RCAF Station Camp Borden.

1939 International Pumper
Two of these pumpers were purchased by the Air Force in 1939. They were the second and third fire trucks owned by the Air Force, and were originally stationed at Camp Borden and Station Trenton. The firefighters, at some point in the service life of the pumper, honoured it with the title of Old Number One.

1941 International Bickle-Seagrave Pumper
Purchased by the Air Force, this vehicle came equipped with a six-cylinder engine and a 625 G/min / 2800 L/min pump at a pressure of 150 psi / 1000 kPa.

1941 Code 30 (Redesignated G10) Pumper
The Code 30 pumper, which was re-designated G10 and become more widely known by the latter, came equipped with an open-cab mounted on an International chassis. It was the principal structural firefighting vehicle of the early post World War II era. It had a rotary-gear positive displacement pump, rated at 600 G/min / 2730 L/min at 120 psi / 825 kPa. The pump was fitted with a spring-loaded, adjustable pilot valve, which in turn controlled a churn valve. During pumping operations, the pilot valve would be adjusted to the desired maximum pressure. When this maximum pressure was exceeded, as might be the case when one or more hand-lines were shut down, the spring in the pilot valve would be compressed, allowing water to flow to the churn valve, causing it to open a bypass between the discharge side of the pump to the intake side of the pump, thereby allowing the pump to churn.

Equipment included a 24 ft / 7 m extension ladder, 1200 ft / 365 m of 2½ inch / 65 mm hose and 300 ft / 90 m of 1½ inch / 38 mm interior attack hose. It also carried ancillary equipment, such as axes, wrenches, pry-bars and extinguishers. It also had a 12 ft / 3.5 m roof ladder, a pike pole and an 80 G / 365 L water tank.

A leftover from the war years, the pumper needed a complete overhaul if it were to remain in service or, failing that, be replaced. With money for the purchase of new vehicles in short supply, the decision was made to modify the vehicle from top to bottom, using available maintenance funds. The chassis was reinforced to accommodate an upgrade that included carrying a 40 ft extension ladder over the hose bay and open-cab. At the Bickle-Seagrave plant in Woodstock, Ontario, the rotary-gear, positive displacement pump was replaced with a centrifugal pump, and the 80 G water tank was replaced by a 300 G / 1365 L tank. In addition, the modification included two low-pressure hose reels and two hydroblenders.[6] The G10 then underwent a major overhaul of its engine and drive train at No. 6 Repair Depot, Trenton, Ontario. While 17 of these vehicles were modified, the project was soon terminated, and an equipment renewal program was introduced to acquire new vehicles.

1941 Code 33 (Redesignated G15) Crash Tender
The G15 built on a Ford chassis that used a Marmon-Harrington drive train was the most widely used airport crash vehicle during the war years. It carried 300 G / 1365 L of water 40 G / 182 L PF and was equipped with a rotary gear pump capable of producing 350 G/min / 1600 L/min of water at 150 psi / 1000 kPa via two 1½ inch / 38 mm discharge ports, each controlled by quarter turn valves located at the rear of the vehicle. Air aspirating play-pipe nozzles,[7] attached to two 100 ft / 30 m hoselines, gave the truck a foam-producing capability. Powered by an 85-horsepower engine, the 1941 model attained a respectable speed of 50 mph / 80 kmh. It had a manual transmission with high/low range and four-wheel drive; the front wheel drive could be disengaged when travelling on firm terrain. An ability to achieve maximum speed quickly became a point of personal and professional pride and much was made of a driver's dexterity. Another interesting feature of this vehicle was the routing of the exhaust system through a box that contained the under-truck piping. The theory advanced was the heat given off by exiting exhaust gas would prevent the water lines from freezing.[8] Although there are no records as to how effective this method was, at the very least it represents

---
[6] See Appendix B for definition.
[7] Ibid., p. B-3.
[8] Major Phil Brown (1947–75), *The Royal Canadian Air Force Fire Service, 1939–1975, A Subjective History*.

a thoughtful, innovative, and inexpensive approach for alleviating a chronic problem.

The need to reach the site of an aircraft crash quickly made increasing vehicle speed an all-consuming goal. In 1942, successful trials were conducted, using an upgraded 95-hp engine, and the vehicle speed was boosted to more than 60 mph / 97 kph. Later, the 95-hp engine became the standard for all trucks produced in 1942. Ancillary equipment included two 100 lb / 45 kg CO_2 extinguishing system. It came complete with two 100 ft / 30 m lengths of 1 in / 25 mm high-pressure discharge hose, mounted on reels on either side of the vehicle between the cab and the body.

1942 American-LaFrance Pumper
Originally built for the City of New York Fire Department (NYFD), the Canadian government obtained the vehicle and placed it in HMC Halifax Dockyard. Officially known as RCN-1000, the nickname "Big Bertha" was given by the firefighters. This vehicle is now restored and located back with Halifax Dockyard Fire Department.

1944 American-LaFrance Aerial
This 75 ft / 23 m aerial was custom-built for the Navy and was numbered 1033.

1945 Gifts from the United States
Shortly after the war ended in 1945, the United States gave Canada a number of surplus ARFFVs that had been positioned along the Alaska Highway. They included two different models, a Class 125 G15 carrying 250 G / 1140 L of water, with a three-piston pump for producing high-pressure fog, and a Class 155 model, with a large open-cab truck carrying 840 G / 3800 L of water. The Class 155 model, with a separate pump engine capable of providing highly pressurized water to both turrets and to three handlines, or by connecting an inductor to one of the turrets, generating foam.

1948 G17 DC Crash Vehicle
The G17 DC crash vehicle was purchased by the Air Force.

Main features: a G15 chassis, four 300 lb / 135 kg units DC, for a total of 1200 lb / 545 kg of DC, and pressurized nitrogen as the expellant.

1951 Chevrolet Range Truck
The Army Fire Marshal obtained four of these range trucks.

Main features: a 1940–41, three-ton 4X4 military pattern vehicle (MPV) chassis, a Hale pump, a 350 G / 1600 L water tank, two 10 ft / 3 m lengths of hard-suction hose, two hydroblenders, and 75 ft / 23 m 25 mm of hose on two hose reels.

1951 G18 DC Vehicle
Assembled for the Air Force, the G18, perhaps, can lay claim to the best or the worst piece of fire apparatus that ever existed, even though it never went into service.

Main features: a F-1000 chassis, a Waukesha GK-145 gasoline engine, a two-stage centrifugal pump rated at 600 G / 2730 L at 150 psi / 1035 kPa, two steel tanks, each with the capacity to contain 2000 lb / 900 kg of DC, eight 400 ft^3 / 11 m^3 cylinders of nitrogen, a 300 G / 1360 L water tank, two turrets, and two reel-mounted handlines.

1951 G9 Pumper
Designed and built by Bickle-Seagrave of Woodstock, Ontario, the G9 triple-combination pumper was the first pumper with an enclosed cab to enter military service.

Main features: a 500 G / 2275 L water tank, 600 G/Min / 2700 L/min two-stage centrifugal pump, later upgraded to 840 G/min / 3800 L/min, the accepted standard for a post-war Class A pumper,[9] and a 40 ft / 12 m extension ladder, hence the Pumper designation.

1952 Bickle-Seagrave Aerial
This 65 ft / 20 m aerial was purchased by the Navy.

Main features: a 12-cylinder gasoline engine, a manual transmission, a hydraulically operated aerial ladder; and a mix of ground ladders totalling 200 ft / 60 m.

1952 Ford Crash Tender
Purchased by the Navy, this was a very versatile vehicle.

Main features: a 400 G / 1800 L water tank; a 30 G / 136 L foam tank, six 100 lb / 45 kg cylinders of CO_2, a pump with a capacity of 250 G/min / 1140 L/min at 120 psi / 850 kPa, the capability of producing foam, a remote-control turret, 200 ft / 60 m of 1 in / 25 mm high-pressure hose, both ground-sweeps and under-truck nozzles, and 200 ft / 60 m of 3/4 in / 20 mm of hose carried on a hose-reel for delivering CO_2.

1952 Pumper
Built in Stratford, Ontario, for the Army and used at Camp Wainwright, Alberta.

Main features: a 255 in^3 / 2.5 L Mercury gasoline engine, and a 500 G/min / 2300 L/min American-Marsh pump.

1952 G11 Pumper
These pumpers were built for the Air Force by Thibault.

Main features: a main-pump having a rated capacity of 840 G/min / 3800 L/min at 150 psi / 1000 kPa, a high-pressure pump capable of producing 800 psi / 375 B, plastic hydroblenders which could not be used with the high-pressure pump, and a mix of hose and ladders.

This vehicle was found to be tail-heavy. As well, this early version of the vehicle had plastic hydroblenders,[10] which could not be used with the high-pressure pump. These problems were corrected in the 1953 model, G11, which then gave the fire services many years of good reliable service.

Bickle-Seagrave and American-LaFrance also built a number of these Class A pumpers which met all requirements given in what had come to be known as the standard Air Force specification. The LaFrance pumpers were exceptionally reliable. One thoughtful feature of the LaFrance pumper was the placement of the pump panel on the curb-side of the vehicle, to allow the operator to keep out of the way of passing traffic.

The positioning of the pump panel became a controversial topic, but the whole question of pump-operator safety was resolved by positioning the pump controls amidships on a raised platform on the new generation of pumpers that came into service in the early 1990s.

The only serious drawback the G11 had was its two-wheel drive chassis which caused some difficulties during winter conditions on some Radar stations where the Operations Buildings were on a hilltop, well above the level of the technical and domestic facilities where the fire hall was located. It was, therefore, generally a steep climb from the fire hall to operations site. This shortcoming was corrected with the delivery of the G8.

1952 G13 LRV
Despite the problems experienced with the ill-fated G18, the more positive aspects of using DC as a fire-

---

9 Today, the standard pumping capacity for a Class A pumper in the US and Canada dictates an output of 840 Imperial gallons per minute.
10 See Appendix B for definition.

extinguishing agent, as was evident with the G17, kept the interest level high. Accordingly, in 1952 the Air Force introduced the G13, a relatively small and fast crash truck that became the Air Force LRV. The reduced weight allowed for incorporation of the desired features of speed and off-road capability. The International chassis for the first model proved to be too light and only ten were produced. A serious design flaw caused the rear axle to break quite frequently. The second version, with a longer manufacturing run, used a Ford 4x4 chassis. The replacement vehicles, with stronger chassis, remained in service for many years.

The fire package was 1000 lb / 450 kg of DC until the early 1960s when PK began to be used in lieu of DC, and two 220 ft^3 / 6 m^3 cylinders of nitrogen. Capable of knocking down a large fire, although not preventing re-ignition of liquid fuels in the event that smouldering combustible material or other hidden fire was in the area. The securing of a fire area of this nature required the application of foam. Overall a serviceable vehicle, it performed a variety of functions for the fire crews.

Over the years, the G13 became somewhat of a utility vehicle. The front-mounted winch performed a myriad of tasks, ranging from releasing aircraft following arrestor barrier engagements, to pulling cars out of ditches. They certainly became especially valuable maintaining and resetting aircraft arrestor barriers on airfield runways after the Fire Service took over this responsibility in the early to mid 1960s. This was an added dimension, not envisaged at the time the vehicles were purchased, that was exploited with much success.

The G13 was a useful addition to a fire department's mobile inventory. In 1964, an updated International chassis replaced the venerable Ford, but did not quite realize the success achieved by its predecessor. Overall, the G13, of which there were several models, had a long and successful life from introduction in 1952 to 1982 when it was replaced with the 2000 L / 440 G Rapid Intervention Vehicle (RIV).

1953 GMC-Bickle-Seagrave Pumper
This pumper was built for the Army.

Main features: a six-cylinder gasoline engine, a standard three-man cab, a 840 G/min / 3800 L/min two-stage pump, a 80 G / 300 L water tank, two hose reels, 150 ft / 45 m of 1 in / 25 mm hose, 1200 ft / 365 m of 2^1/$_2$ in / 65 mm hose, a 24 ft / 7 m and a 14 ft / 4 m ladder. It is not known how many of these trucks were made or where they were allocated.

1953 G21 Thornycroft MFV
In 1953, the Thornycroft Company of England produced a MFV, designated G21.

Main features: a 400 G / 1800 L water tank, a 100 G / 450 L foam tank, an eccentric rotary vane pump, a foam-generating system using 6% PF to achieve a 14:1 expansion ratio, a turret, along with pump controls, mounted on an upper rear-facing platform, and a right-hand drive.

The eccentric rotary-vane pump, which was new to most firefighters, required frequent maintenance. As well, the turret's rear-facing position required a change of tactics by the vehicle operators, long accustomed to a forward-facing turret with a straightforward head-on approach to an accident scene. They now had to adapt to a swing-away approach that left the vehicle facing away from the incident. Considerable judgment on the part of the driver was required to arrive at the scene of the incident, negotiate an 180° about-face without interfering with, or striking, another vehicle in order to arrive at a position facing away from the target and yet be within effective range of the turret!

To further complicate matters, with the driver's position located on the right side of the vehicle, in keeping with its country of origin, drivers had to change gears with their left hand, a switch that did not take too long to master. Perhaps a greater challenge faced firefighters stationed at No. 30 Air Materiel Base in Langar, England, and No.1 Fighter Wing in North Luffenham, England, who had to adapt to driving on the left-hand side of the road. This feat generally took a little longer to master. Many firefighters were given a driving adaptation course at RAF Station Wheaton (near Blackpool), England.

It was specifically built to operate in the crash-protection role and saw service only on Canada's NATO airfields in Europe. Every Canadian airfield in England, France and Germany took delivery of at least one of these vehicles in the early 1950s.

1953 A21 FWD MFV
Built for the Air Force, This vehicle represented a significant step forward in crash vehicle technology.

Main features: a custom FWD 4X4 chassis, a 240-hp Waukesha gasoline engine, a 500 G / 2275 L water tank, 70 G / 320 L of 6% PF concentrate, a separate pump engine, a roof-mounted dual-stream turret, a hose-reel on each side with 100 ft / 30 m of 1 in / 2.5 mm hard-wall rubber-hose, the capability to deliver foam at the rate of 500 G/min / 2275 L/min with a reach of stream of up to 85 ft and an expansion ratio of 8:1.

This truck could be used on structural fires by closing a valve on the foam line and using the hand lines with water only. The arrival of this vehicle allowed the G15 to be phased out. Up to this point all new crash vehicles had something in common—they all had limited off-road capability. For reasons that could not be determined, the A21 was only in use a very short time before it was replaced with a G23 FWD built in 1954.

1954 G23 FWD MFV
This MFV gave the fire departments a reasonably effective vehicle for combatting flammable liquid fires. As such, it became the primary ARFFV.

Main features: a custom FWD 4X4 chassis, a 240-hp Waukesha gasoline engine, a 500 G/min / 2275 L/min centrifugal pump directly coupled to a 163 hp gasoline engine, a 500 G / 2275 L water tank and a 70 G / 320 L PF tank, a roof-mounted turret, two ground-sweep nozzles under the front bumper, two one-inch hand-lines mounted on reels on each side of the vehicle complete with three position nozzles and two under-truck nozzles to protect the underside of the vehicle. The agent of choice was PF and, although the foam did not have the same rich white appearance of later protein foams, it was, nevertheless, very effective in the hands of a skilled operator.

The truck was highly manoeuvrable on hard surfaces, but unfortunately lacked a legitimate off-road capability. The turret operator stood on the outside of the vehicle and communicated to the driver through an intercom system. The position was one of good advantage and provided excellent all-round visibility. The turret operator had complete control of the pump engine and foam-delivery system. The turret operator used a governor-controlled air switch to bring the Chrysler 24A industrial V-8 gasoline pump engine up to speed and the pump pressure to pre-set values. Features, such as the dual-type turret that could discharge foam or water as a fog or straight-stream and handlines that were easy to deploy, allowed this vehicle to remain in service until the early 1970s.

In a secondary role, this vehicle was also used for fighting structural fires. The technique involved closing the valve between the pump discharge and foam-proportioned water, so that straight water could then be used through the turret and hand lines. Water could also be taken on from a hydrant or other pressure source through a pair of 2½ in / 65 mm ports. However, there was no provision to draft water from a static source.

Driving the G23 could be a traumatic experience. With five very closely spaced gears in low range and a further five in high range, this was not the vehicle for the timid or unco-ordinated. Overall, the G23 was a good solid vehicle. The most serious drawbacks to the vehicle's performance were its modest water-carrying capacity, low foam-expansion ratio, and limited off-road capability.

1954 GMC-Bickle-Seagrave Pumper
This pumper was built for the Army.

Main features: a GMC commercial chassis, a six-man cab, a 500-hp engine, a two-stage Hale 840 G/min / 3800 L/min, a 300 G / 2300 L water tank, two 1½ in / 38 mm hose reels connected to a booster pump, and two hydroblenders.

1954 FWD Pumper
This vehicle, a Class A pumper with 840 G/min / 3800 L pump capacity, was used by the Navy at a number of their units.

1954 Bickle-Seagrave Pumper
This pumper was purchased by the Navy.

Main features: a 500-hp, 12-cylinder gasoline engine, a parallel-series centrifugal 840 G/min / 3800 L/min pump, a 300 G / 1365 L water tank, and space for three firefighters to ride inside the cab.

1954 Walter, Bickle-Seagrave MFV
This MFV was built for the Navy by Maxim Motors in the US.

Main features: a custom chassis was by Walter, a 240-hp Waukesha gasoline engine, a pair of 600 G/min / 2700 L/min pumps powered by two Chrysler IND-9 Industrial engines, rated at 140 hp, a 840 G / 3800 L water tank, a 75 G / 340 L PF tank, twin 165 G/min / 750 L/min turrets, two 100 ft / 30 m handlines, and 800 lb / 360 kg of CO_2, which discharged through a fibre horn.

1955 International-Thibault Pumper
This pumper was built for the Navy.

Main features: an International commercial chassis, a six- or eight-cylinder engine, a manual transmission, a 300 G / 1365 L water tank, a parallel-series centrifugal pump with a capacity of 840 G/min / 3800 L/min, and a cab with space for five firefighters to ride inside this vehicle.

1956 International Harvester Crash Truck
The Navy purchased this pick-up style truck to carry a skid-mounted ARFF unit.

Main features: an International 4X4 pick-up style truck, a six-cylinder engine, a 7 G / 30 L foam tank and a 110 G / 450 L water tank, a 250 ft$_3$ / 7 m$_3$ air cylinder with a working pressure of 400 psi / 2800 kPa to discharge the agent at 25 G/min / 115 L/min at 120 psi / 850 kPa, and two 50 ft / 15 m hose reels. The whole mounted on a skid-pack.

1957 G19 Alvis MFV
The 1957 Alvis G19 MFV was manufactured in England, for the Air Force, by the Pyrene Foam Company.

Main features: a 6x6, all-wheel drive Salamander chassis, which had been designed for use as an armoured personnel carrier, a Rolls-Royce B81, Mk 80A in-line, eight-cylinder engine, providing motive power as well as power to the pump through a power take-off, a pre-select transmission, a driver's position in the centre of the cab, 700 G / 3200 L of water and 100 G / 455 L of 6% PF concentrate, a 800 ft^3 / 23 m^3 per minute air-blower, a roof mounted turret, and two 4 in / 10 cm hand-lines.

One of the more distinctive features of this vehicle was the pre-select transmission. It required the driver to pre-select the desired gear, move the gear selection lever to that position, and when ready to make the change, quickly depress and release a foot-operated gear-change pedal. The transmission was overtaxed in this role due to the weight/mass of the firefighting package and was prone to breakdowns. As a result, the engine and transmission assembly was removed and the V-8 Ford engine and Allison automatic transmission was installed. Despite these changes, the vehicle had a short lifespan and began to be replaced in the mid 1960s, although some remained in service into the 1970s.

The Pyrene foam-producing system, utilizing an air blower, increased the foam expansion ratio, producing foam with exceptional cohesion and adhesion qualities. All controls for operation of the foam-producing equipment were located in the cab, and delivery valves to handlines and turret could be operated from the driver's seat, which was located in the centre of the cab, another of this vehicle's unique features. The turret was exceptional, with an effective reach of stream of over 100 ft / 30 m. When the G19 Alvis was serviceable, it was an unmatched piece of firefighting equipment. However, the four-inch handlines did become extremely heavy when charged with foam. To manoeuvre charged lines was a real test of strength.

Its cross-country ability was remarkable, and it came very close to being the much sought-after all-terrain fire vehicle. However, the payload was heavier than the chassis was meant to carry. As a result, the vehicle became beset with mechanical problems and quickly earned the title of "Hangar Queen." The title, which it garnered in the early stages of its period of service, may not have been totally deserved. The vehicle's design was basically sound, and its cross-country, off-road performance was outstanding. The off-road capability of this vehicle has yet to be matched by newer designs. The G19 Alvis raised the bar in the MFV category.

1957 G8 PUMPER
This pumper purchased by the Air Force ushered in a new generation of structural firefighting vehicles. Built by the Pierre Thibault Company, the G8 was a composite of fire equipment and vehicle parts from a variety of sources, as is often the case with fire apparatus. The combination in this case produced a thoroughbred.

Main features: a 4x4 design chassis by FWD, a Waukesha straight eight engine with a dual 24-volt ignition system, a Waterous Class A 840 G/min / 3800 L/min two-stage, centrifugal pump, a 500 G / 2275 L water tank, a four-stage, high-pressure pump that was connected to two high-pressure hose reels located on either side of the vehicle, a hydroblender system, incorporated into the high-pressure unit, 1200 ft / 365 m of 2^1/$_2$ in / 65 mm hose, 300 feet / 100 m of 1^1/$_2$ in / 38 mm hose, a 12 ft / 3.5 m roof ladder, a folding step/attic ladder, a 40 ft / 12 m and a 24 ft / 7 m extension ladder.

By increasing the penetration of water into small, deep-seated fires in mattresses, upholstered furniture and similar material, hydroblenders were effective. Unfortunately, the hydroblender agent that created the wet-water action clogged lines and the benefits over using plain water were placed in question. Eventually the hydroblender unit was scrapped, and subsequent pumpers did not have the high-pressure pump, hose reels or the hydroblender feature.

The rationale for the removal of the high-pressure/ hydroblender combination was due in part to serviceability problems and to the belief by some authorities (not only military) that fire crews, upon arriving at the scene of a fire, would commonly employ the small, high-pressure lines in an almost automatic reaction. These lines, although light and quick to bring into action, invariably could not produce the volume of water necessary to control sizable fires. This occasionally resulted in some fires getting out of control. The consensus among experts in firefighting tactics was that the minimum size hose for effective initial attack was 1½ in / 38 mm hoselines fed from the main pump. Universal acceptance of this principle effectively killed the high-pressure pump/hydroblender concept.

The G8 carried a foam inductor designed to be used directly connected to a 2½ in / 65 mm discharge port with a pick-up hose and screened metal probe designed to take foam concentrate directly from a 5 G / 23 L container of foam concentrate. This enabled the G8 to fight medium-sized, flammable-liquid fires or provide a secure vapour seal over fuel spills. Extension of this idea led to incorporating an integral foam tank on pumpers beginning in the mid 1970s.

By incorporating a simple, virtually trouble-free pump operation, mechanical breakdown was rare. Overall, it was a reliable piece of fire apparatus.

The High-Pressure Pump
The high-pressure pump and hydroblenders features of the G11 and the G8 pumpers had, in addition to the main pump, discharged water through a high-pressure hose, which was mounted on hose reels. The nozzles were of a Hardie gun-type design with a tip approximately 0.5 in / 13 mm in diameter. The pump, which produced 800 psi / 375 B, was a four-stage centrifugal design in a cylindrical shape. It was driven by a power take-off and ran at relatively high revolutions per minute (RPM), a combination that emitted a confidence-busting, high-pitched sound, quite like a scream. Not surprisingly, the system proved to be unreliable.

1958 Ford Range Truck
This vehicle was built for the Army.

Main features: a three-ton, 4x4 MPV chassis, a Mercury gasoline engine, a Wajax Mark 1 centrifugal high-pressure four-stage pump, a 2 in / 50 mm hard-suction hose, and a 1½ in / 38 mm discharge port.

1958[11] E62 Street Flusher
Another vehicle that the fire chief at airports had in the arsenal of firefighting equipment was the E62 Flusher. As the name suggests, this vehicle's primary role was to clean streets, ramps and other hard surfaces. However, some far-sighted individual decided to equip the E62 with foam-making capabilities, making it a versatile auxiliary fire vehicle.

Over time, a number of aircraft emergencies involving defective landing gear had occurred at Canadian airports and those of other NATO countries. In some instances, foam paths, using vehicles similar to the E62, had been laid prior to the aircraft landing with good results. The successes had allowed the practice of laying runway foam paths to become a recognized method of dealing with certain types of aircraft emergencies. A request to foam a runway almost invariably came from the aircraft commander.

The E62 could lay a foam path 9 ft / 2.7 m wide, 2000 ft / 610 m long and 2 in / 5 cm deep at a speed of 4 to 5 mph / 6 to 8 kph. The foam was applied through a manifold arrangement attached to the rear of the

---

[11] Date not confirmed.

vehicle that was fitted with a number of nozzles. There appeared to be at least four possible benefits to be gained from foaming a runway: reduction of aircraft damage; reduction in decelerating forces, reduction of friction-spark hazards,[12] and reduction in fuel-spill hazards. The theory upon which the practice was based was that the film of water that would leach out of the foam blanket would also be held in place by the foam bubbles, and that this water would eliminate sparking as a source of ignition.

PF generally had a higher expansion ratio and a longer drainage time[13] than synthetic foam. It could, therefore, remain in a condition that supported the theory for a sufficient time that, on the one hand, allowed time for the application process and, on the other hand, to get the aircraft down.

Analysis of the theoretical benefits of foaming runways remained inconclusive; however, with the introduction of AFFF foam, the practice ceased.

Despite the loss of its runway-foaming role, the E62 continued to provide the firefighter with an important firefighting asset. It could readily be used to supply water to nurse crash vehicles or employ its 2000 G / 900 L water tank to contribute to fire department operations in outlying areas. The Flusher was also used for decontamination operations at nuclear-capable bases and at bases that maintained a team for dealing with nuclear accidents.

1960 Fargo Aerial
This 65 ft / 20 m aerial was used by the Navy at HMC Halifax Dockyard.

1960 Tracked DC Crash Vehicle
The Air Force purchased this tracked vehicle for service in Resolute Bay, Northwest Territories.

Main features: a chassis mounted on tracks, a 351 in^3 / 3.5 L Ford V-8 engine, a 500 lb / 225 kg DC unit, and two handlines. It could carry 12 personnel, or be fitted to carry four stretchers when needed to rescue stranded people.

1962 Ford Range Truck
This vehicle was built for the Army.

Main features: a 4x4 MPV chassis and an eight-cylinder gasoline engine.

1963 MPV Pumper
The Navy purchased six of these pumpers.

Main features: a 1958 2^1/$_2$ ton 6x6 MPV chassis, a four-stage 80 G/min / 360 L/min high-pressure pump at 200 psi / 1380 kPa, a 800 G / 3600 L water tank, two 200 ft / 60 m hose reels, two hydroblenders for blending wet-water, and a portable Wajax forestry pump.

1964 Thibault Pumper
This pumper was built for the Army.

1964 G19 Sicard MFV
This MFV was built for the Air Force.

Main Features: a Sicard 6X4 chassis with tandem single-wheel axles, set towards the rear, a Waukesha V-8 prime mover engine, a Rolls-Royce pump engine in vehicles built in 1964, a Ford V-8 pump engine for vehicles built after 1964, an Allison automatic transmission, a Pyrene foam-producing system for inducting PF concentrate at a 6% rate and, utilizing an 800 ft^3 / 23 m^3 per minute air-blower, achieve a 14:1 expansion ratio, a roof-mounted turret, and two 100 ft, 4 in. / 10 cm hand-lines.

---

[12] Resulting from the friction of the aircraft's metal components rubbing against the runway surface.
[13] The time it takes for a given percent of liquid to leach out of the expanded foam.

Designed to deliver a formidable 5000 G/min / 22700 L/min of expanded foam to the turret in its high-rate setting or 2500 G/min / 11,350 L/min to the turret in its low-rate setting, with the remaining capacity directed to hand-lines or other outlets. The turret with a maximum range in the neighbourhood of 120 ft / 35 m, varied at the discretion of the operator down to 35 ft / 11 m by using a fan mode of operation. The handlines, conveniently stored on each side of the vehicle, were extremely difficult to manoeuvre, once charged.

A very reliable vehicle, it remained in service into the late 1970s. It was the last MFV purchased that used PF.

1965 Ford, King-Seagrave Pumper
This pumper was purchased by the Navy.
Main features: a Ford F500 series truck frame, a manual transmission, a Hale 420 G/min / 1900 L/min single-stage pump, and a 300 G / 1365 L water tank.

1965 Walter, King-Seagrave MFV
This MFV was built for the Navy.

Main features: a 534 in^3 / 8.75 L Ford gasoline engine, giving it a max speed of 60 mph in 45 seconds / 97 km in 45 seconds, a 840 G / 3800 L water tank, and a 90 G / 410 L PF tank.

1966 Thibault Pumper
This pumper was built for the Army.

Main features: a gasoline engine, an electric assist manual transmission, an 840 G/min / 3800 L/min pump, and a 500 G / 2300 L water tank.

1969 Heliport ARFF Vehicle
This vehicle was produced by the Pyrene Manufacturing Company to meet the Canadian Forces requirement for a self-contained, air-transportable firefighting package for crash protection for the Tactical Helicopter Squadrons with the Army Brigade Groups at CFBs Valcartier, Petawawa, and Edmonton, and the training school in CFB Gagetown.

Main features: an International Harvester Model 1700 Loadstar series with 4x4 drive chassis, an automatic transmission, a lightweight twinned-ball unit containing 400 lb / 180 kg of PK and 50 G / 225 L of pre-mixed water and AFFF, two, reel-mounted, handlines, and nitrogen cylinders.

The unit was rated to extinguish and secure a flammable liquid fire, with a surface area of up to 2000 ft^2 / 185 m^2. The prototype was delivered to CFB Petawawa in January 1969.

1970 UNIMOG Quick Response Vehicle
This vehicle, which carried both AFFF and PK, a practice that was to be widely accepted in future years, was purchased in Europe by the air force units in West Germany.

Main features: a Mercedes UNIMOG 4x4 chassis, a Mercedes diesel engine, a pressurized tank with a capacity of 1000 L / 220 G of pre-mixed AFFF/water solution, a similar tank with a capacity of 225 kg / 500 lb of PK, nitrogen for expelling for both agents, a turret and two handlines. Either agent could be applied singularly or in combination with the other.

1972 International-Pierreville Pumper
Equipped to Canadian Armed Forces standards, its main features a include a 840 G/min / 3800 L at 150 psi / 1000 kPa pump and a 500 G / 2300 L water tank.

1972 Oshkosh M-1000 MFV
The Oshkosh M-1000 and the T-1000, essentially purchased at the same time, were the first MFVs designed to use AFFF. Both were built by the Oshkosh Company, Oshkosh, Wisconsin. Over time, they

replaced the G19 Alvis and the G19 Sicard and ended the use of PF.

Main features: a Oshkosh 4X4 chassis, a Detroit diesel engine, an Alllison automatic transmission, a pump capable of developing a pressure of 250 psi / 1700 kPa, a 840 G / 3800 L water tank, a 120 G / 550 L foam tank, twin turrets, two handlines, under-truck nozzles, and front-mounted nozzles.

Capable of reaching a speed of 50 mph / 80 kmh within 60 seconds, the M-1000 vehicles were positioned originally in Baden-Sollingen, West Germany, the Fire School in Camp Borden, and CFB Bagotville.

1972 Oshkosh T-1000 MFV

The T-1000 series had the same specifications as the M-1000, with the exception that the M-1000 vehicles had twin turrets and the T-1000 had a single turret.

1974 MAN Gebruder Bachert Pumper

These vehicles were purchased in Germany and used at both CFB Lahr and CFB Baden-Sollingen.

Main features: a MAN commercial chassis, a MAN six-cylinder diesel engine, capable of accelerating the vehicles to a speed of 80 kmh / 50 mph with ease, a manual transmission, a 3800 L/min / 840 G/min pump, a deck-mounted nozzle, a built-in foam-induction system, a 2275 L / 500 G water tank, and a 300 L / 60 G MXF tank.

These vehicles were well-made, low-maintenance trucks that served the Forces well. When the bases closed, these vehicles were the only MAN pumpers in Europe and were placed in museums.

1974 Ford Saskatoon Barton-American Pumper

These vehicles were built in Calgary, Alberta, by the Saskatoon Fire Engine Co., strange as that may seem.

Oshkosh M-1000

Major Phil Brown tells a story of how the CF happened to come by the Oshkosh M-1000:[14]

It began while I was attending a meeting of the NFPA Foam Committee held in Ottawa, in 1972. At the same time we received a FLASH message from No. 1 Air Division in Lahr, West Germany, that two of the three MFVs at No. 4 Wing, Baden-Sollingen were down for lengthy repairs and that flying would be curtailed until ARFF services was brought up to strength. Remember this was at the height of the Cold War.

During the meeting, I happened to mention that it was too bad MFVs were not an off-the-shelf item, as we needed some urgently. Another committee member piped up and said his company had several that we could buy immediately. The chap was the sales manager for Oshkosh. I phoned Lieutenant Colonel Chisholm, who was the CFFM at that time. He came up to the meeting that afternoon and we were soon closeted with the sales manager.

It turned out that Oshkosh was a major manufacturer for the US Navy which had ordered 175 crash trucks and had cancelled part of the order after receiving 100 crash trucks. Oshkosh had several that had been completed when the US Navy downsized their order. Two could be ready for shipment in a week. You can imagine our joy. Lieutenant Colonel Chisholm swung into action. He obtained authority to purchase two on an emergency basis. He had all kinds of help from the Director of Flight Safety, Operations, and several other stakeholders.

The purchase order was authorized, and two Hercules C-130 aircraft transported the vehicles from Oshkosh to Lahr. The whole exercise was completed in three weeks.

---

14 Major Phil Brown (1947–75), *The Royal Canadian Air Force Fire Service 1939–75, a Subjective History.*

Main features: a Ford commercial chassis, a V-8 225-hp Caterpillar diesel engine, giving the vehicle a maximum speed 47 mph / 75 kmh, a Barton-American 840 G/min / 3800 L/min pump, and a 500 G / 2300 L water tank.

1975 International King-Seagrave Pumper
Main features: an International chassis, a turbo-charged V-6 Detroit diesel, capable of taking the vehicle to a speed of 55 mph / 90 kmh, a 840 G/min / 3800 L/min pump, a 500 G / 2300 L water tank, and a high-volume deck-gun mounted above the pump compartment.

1975 Ford Aerial
Main features: a Ford chassis, a 636 in^3 / 10.4 L Caterpillar diesel engine, capable of taking the vehicle to a speed of 50 mph / 80 kmh, a 75 ft / 23 m aerial ladder, and an 840 G/min/ 3800 L/min pump.

1976 FLEXTRAC PK Crash Vehicle
This fully tracked, crash fire vehicle was purchased for service at CFS Alert, Northwest Territories. It was prone to mechanical failure and was, at times, derogatively referred to as the "Hangar Queen." The firefighting package was 1000 lb / 450 kg of PK using pressurized nitrogen as the expellant.

1977 Kenworth Pierre Thibault Aerial
Main features: This 65 ft / 20 m aerial had a Kenworth chassis, a V8 Detroit Diesel and a Waterous 840 G/min / 3800 L/min pump.

1978 Scot, Pierre Thibault Aerial
Main features: a custom Pierre Thibault chassis, a 65 ft or 75 ft / 20 m or 23 m ladder, either a V-6 or a V-8 Detroit diesel, depending on the size of the ladder, with a top speed of 60 mph / 100 kmh, and an 840 G/min / 3800 L/min pump.

1979 IHC Pierre Thibault Pumper
This pumper was designed, specifically, to climb the hills normally found around radar sites.

Main features: a 466 in^3 / 7.6 L diesel engine, a front-mounted 420 G/min / 1900 L/min pump, a 600 G / 2700 L water tank and a 40 G / 180 L foam tank.

1979 Oshkosh M-1000 MFV
This vehicle replaced the last of the Sicard MFVs.

Main features: a custom Oshkosh chassis, a six-cylinder Caterpillar diesel engine that could accelerate from 0–80 kph / 0–50 mph in 35 seconds, with a top speed of 60 mph / 96 kph, a V-8 90-degree Caterpillar diesel pump engine, a 840 G / 3800 L water tank, a 110 G / 500 L foam tank, a hydraulic, remotely controlled turret, a 225 kg / 500 lb PK unit, a Halon 1211 wheeled fire extinguisher with a capacity of 80 lb / 36 kg, two handlines, and both ground-sweep and under-truck nozzles.

1981 International Pierreville Aerial
This vehicle was used in CFB Lahr.

Main features: an International commercial chassis, a 100 ft / 30 m aerial ladder, a V-6 Detroit diesel, giving the vehicle a top speed of 55 mph / 88 kph, a Waterous 1100 G/min / 5000 L/min pump, a 440 G / 2000 L water tank, and a 40 G / 180 L MXF tank.

Increased Demands—New Solutions
Larger aircraft demanded larger and faster fire vehicles to cope with increased fuel and passenger loads. The acceleration for major fire vehicles was increased from an accepted 0–50 mph / 0–80 kph in less than 60 seconds, to a new criterion of 0–50 mph / 0–80 kph in less than 20 seconds. As well, vehicle firefighting payloads were increased dramatically.

1981 RIV
The 1980s was the beginning of major changes in fire protection equipment, and especially in ARFFV. ARFF standards were changing, based on data collected over

time, and were calling for more foam and water with a faster response time. To meet the challenge, the military developed the first RIV in 1981, built in Montreal by Walters.

Main features: a custom 4X4 chassis, a turbocharged Detroit diesel V8 engine, capable of accelerating the vehicle from 0–80 kph / 0–50 mph in 18 seconds, a 2000 L / 440 G water tank, a 250 L / 50 G AFFF tank, a 225 kg / 500 lb PK unit, a 6 m^3 / 220 ft^3 cylinder of high-pressure nitrogen as the energy source to discharge the PK, a coaxial dual-agent turret for applying either agent individually or in combination, and a PK handline. It was capable of discharging its 2120 L / 465 G of foam/water mix in 75 seconds.

The RIV replaced the long-serving G13. It is of interest to note that all ARFFV from this point onwards were fitted with a PK unit to deal with fire in inaccessible places and three-dimensional fires in support of foam, the main fire-suppression agent.

1981 CDN Research MFV

CDN Research of Toronto built six of these MFVs for service at CFB Ottawa and CFB Trenton. The company assigned the name Foam Boss to the vehicles.

Main features: a custom 4X4 chassis, a 492-hp diesel engine, to accelerate the vehicle to a speed of 105 kph / 65 mph, a 4500 L/min / 990 G/min pump, a 6000 L / 1300 G water tank, a 750 L / 155 G AFFF tank, a roof-mounted turret, and two handlines.

It is interesting to note that in respect to the quantity of water carried by the three Foam Boss MFVs and one RIV at both airports, it fell 2290 L / 500 G short of the 22900 L / 5000 G called for in NFPA 403 for a Category 8 airport; however, it was major improvement from years gone by.

1982 Ford King-Seagrave Pumper

Main features: a Ford commercial chassis, a Detroit diesel engine, to accelerate the vehicle to a speed of 106 kmh / 66 mph, a 3800 L/min / 840 G/min, a 2300 L / 505 G water tank, a 180 L / 40 G MXF tank, and a built-in foam inductor, mounted above the pump.

1982 MPV Range-Truck

The Almonte Fire Truck Company of Almonte, Ontario, assembled this unit on a removable skid for mounting on an MPV.

Main features: a 6x6 wheeled MPV chassis, a V8 Detroit diesel engine, a removable skid as a base for the firefighting package, a 2250 L / 495 G water tank, a 180 L / 40 G tank for Class A foam, a Wajax Mark 3 pump, two medium-expansion Class A foam nozzles, and a selection of forestry equipment.

1982 Walter 4500 MFV

This vehicle was built by the Walter Company of Albany, New York.

Main features: a 4X4 Walter custom chassis, a diesel engine, a 3800 L / 840 G water tank, a 540 L / 120 G foam tank, a pump rated at 2800 L/min / 625 G/min, a turret, and two handlines. The turret discharge rate was 2000 L/min / 440 G/min on high rate and 1000 L/min / 220 G/min on low rate. Significantly, this one-of a-kind vehicle did not have a DC or PK system.

At the time of purchase, foam trucks were in extremely short supply, due to the poor serviceability rate of aged vehicles. In this regard, the situation at North Bay was particularly critical. Working in co-operation with the Vehicle Engineering Directorate, CFFM staff explored the market and located an MFV ready for delivery. Funding was obtained and the decision was made to buy the vehicle and send it to North Bay.

1982 Universal Go-Track Crash Vehicle

This vehicle was basically a personnel carrier modified at CFB Portage LaPrairie, Manitoba, to carry a small twin-agent package. It was used at the helicopter Power Failure Landing and Auto-Rotation field Grabber Green. Because the field was rolled snow in winter, which is not navigable with a wheeled vehicle, a tracked vehicle was a mandatory requirement.

1984 Western Star Range Truck

Main features: a Western Star 4X4 chassis, a 210-hp Caterpillar diesel engine to accelerate the vehicle to 88 kmh / 55 mph, a 340 L/min / 90 G/min pump, a 4270 L / 940 G water tank, a 270 L / 60 G Class A foam tank, and two handlines.

1984 Ford Thibault Pumper

Main features: a Ford commercial chassis, a 210-hp Caterpillar diesel engine to accelerate the vehicle to a speed of 95 kmh / 60 mph, a 3800 L/min / 840 G/min pump, a 2300 L / 500 G water tank, and a 180 L / 40 G foam tank.

A new feature on this truck was the 38 mm / 1½ in hose storage bay located above the pump compartment, making it accessible from both sides.

1985 Walter 4500 MFV

This MFV was built by the Walter Company of Montreal.

Main features: a Walter custom 4X4 chassis, a turbocharged V8 Detroit diesel engine, a eight-cylinder engine pump engine, a pump with a capacity of 3200 L/min / 700 G/min, a 4500 L / 1000 G water tank, two foam tanks with a combined capacity of 700 L / 150 G, a 225 kg / 500 lb PK unit, with a flow rate of 6.8 kg/S / 15 lb/S from the turret and 3.5kg/S / 8 lb/S from the hoseline, an electric remote-controlled coaxial dual agent turret, with a flow of 2000 L/min / 440 G/min on high rate and 1000 L/min / 220 G/min on low rate, with a reach of stream of 12 m / 50 ft depending on the flow rate, and a 30 m / 100 ft hoseline with a discharge rate of 250 L/min / 55 G/min.

The vehicle was capable of accelerating from 0–80 kph / 0–50 mph in 24 seconds, and reaching a maximum speed of 117 kph / 73 mph.

1987 E-ONE TITAN MFV

With the purchase of this TITAN from the E-ONE Company of Ocala, Florida, the trend towards larger vehicles appears to have peaked.

Main Features: an E-ONE custom 6X6 chassis, a diesel engine, a centrifugal pump, a 10000 L / 2200 G water tank, a 2000 L / 440 G AFFF tank, a turret with a discharge rate of 5000 L/min / 1100 G/min on high rate and 2500 L/min / 550 G/min on low rate, a 225 kg / 500 lb unit PK unit, and two handlines.

A year later in 1988 the second E-ONE TITAN, on an E-ONE custom 4X4 chassis, was purchased with water and foam capacities reduced to 5900 L / 1300 G of water and 740 L / 160 G of foam concentrate, respectively. The turret discharge rates remained about the same.

1988 Ford Superior Pumper

Main Features: a Ford commercial chassis, a 636 in^3 / 10.4 L Caterpillar diesel engine capable of a top speed of 100 kph / 60 mph, a 3800 L/min / 840 G/min Hale pump, and a high-volume, deck gun mounted above the pump compartment.

1988 Unimog Range Truck

This truck was built for service with the range control group at CFB Suffield.

Main Features: a Unimog 4X4 chassis, a 214-hp turbocharged diesel engine, a pump with a capacity of

680 L / 150 G at 690 kPa / 100 psi, a diesel pump engine, a 1200 L / 265 G water tank, and a 80 L / 18 G foam tank.

1989 Timberjack Range Truck
This truck was built by Timberjack Inc. and Sage Hill Rural Development Corporation Ltd.

Main Features: a Timberjack 4X4 chassis, a 100-hp four-cylinder Cummins diesel engine, a Briggs & Stratton high-torque belt-driven pump with a capacity of 360 L/min / 80 G/min, a 4500 L / 1000 G water tank, and a blade on the front for use in wild-land fire situations.

1989 Waltek Universal Go-Track Crash Truck
This truck was built by the Walter Company of Montreal.

Main features: a full-track chassis, a V6 360-hp Detroit diesel, providing a maximum speed of 45 kmh / 28 mph, an American Godiva pump with a capacity of 2300 L/min / 505 G/min at a 1000 kPa / 140 psi, a 2750 L / 600 G water tank, a 475 L /105 G foam tank, a 225 kg / 500 lb PK system, pressurized by nitrogen, and a dual turret with a discharge rate of 2000 L/min / 440 G/min with foam/water stream and 6.8 kg / 15 lb per minute discharge rate with PK.

1993 Spartan-Thibault Pumper
This Pumper was manufactured by Nova-Quintech, which had purchased the assets of Camions Pierre Thibault after it declared bankruptcy in 1990.

Main features: a Spartan-Thibault custom chassis, a turbocharged Detroit diesel, providing a cruising speed of 100 kph / 60 mph. It had a 4800 L/min / 1050 G/min pump, a 2300 L/ 505 G water tank, twin 32 L / 14 G tanks, a high-volume nozzle mounted on top of the truck and a hose storage-bay above the pump compartment, enclosing a crew-cab and a pump control panel positioned behind the cab. The latter two features enhanced crew safety.

1993 Oshkosh 6000 MFV
This MFV was purchased from the Oshkosh Company of Oshkosh, Wisconsin.

Main features: a custom Oshkosh 4X4 chassis, a V8 Detroit diesel engine, a 6000 L / 1300 G water tank, a 750 L / 165 G foam tank, a pump capable of delivering 5600 L/min / 1230 G/min, a 225 kg / 500 lb PK unit, a coaxial turret for applying either agent individually or in combination, a bumper turret, two handlines, and two under-truck nozzles.

A second model was equipped with a snozzle[15] instead of a turret. The snozzle was mounted on an articulating arm and was able to puncture the fuselage of an aircraft and apply extinguishing agents to the interior.

1995 Navistar-International Range Truck
Main features: a International 4X4 chassis, a six-cylinder turbocharged diesel engine, a 4500 L / 1000 G water tank, plus a 125 L / 28 G foam tank, and a Wajax pump with a pumping capacity of 230 L/min / 50 G/min. It could carry its 5-ton / 4.5-tonne payload with ease while cruising at 80 kph / 50 mph.

1995 Oshkosh T-1000 MFV
This MFV was purchased from the Oshkosh Company of Oshkosh, Wisconsin.

Main features: a custom Oshkosh 4X4 chassis, a V-8 Detroit diesel engine, an Allison torque converter transmission with a power-divider allowing the vehicle a pump-and-roll during firefighting operations, a Waterous pump with a capacity of 3800 L/min / 840 G/min at

---
15 See Appendix B for definition.

1600 kPa / 235 psi, a 3800 L / 840 G water tank, a 530 L / 116 G foam tank, a 225 kg / 500 lb PK system, a dual-agent coaxial turret, and a bumper turret.

It was designed to be transportable in Hercules C-130 aircraft.

1995 Emergency One P-150 MFV
This MFV was purchased from the E-ONE Company of Ocala, Florida.

Main features: an E-ONE custom 4X4 chassis, a V-8 Detroit diesel engine, providing a cruising speed of 110 kph / 68 mph, a pump with a capacity 4700 L/min / 1030 G/min, a 6000 L / 1300 G water tank, a 770 L / 170 G foam tank, a 250 kg / 550 lb PK unit, a coaxial turret capable of delivering foam and PK singularly or in combination, and two handlines.

1995 Tibotrac Water Tower/Pumper
Built by Tibotrac Inc., it was the first water tower purchased by DND.

Main features: a custom 6X4 Tibotrac chassis, a six-cylinder Detroit diesel providing a speed of 100 kph / 60 mph, a 5700 L/min / 1250 G/min pump and, depending on the locations need, a 20 m or 23 m / 50 ft or 75 ft water tower.

The power to raise and lower the tower was provided by both electrical and hydraulic means. As things turned out, both methods experienced many problems, resulting in these vehicles being converted to pumpers.

1995 Oshkosh T-1000C MFV
This MFV was purchased from the Oshkosh Company of Oshkosh, Wisconsin.

Main features: an Oshkosh custom 4X4 chassis, a diesel engine, an Allison automatic transmission with a power divider, providing a pump-and-roll option during firefighting operations, a pump rated at 3800 L/min / 840 G/min, a 3800 L / 840 G water tank, a 530 L / 116 G foam tank, a 225 kg / 500 lb PK unit, a turret, two handlines and under-truck nozzles.

1996 Ford Rescue Van
The CFB Borden Fire Department acquired this Ford van to carry specialized equipment in support of firefighting and rescue operations.

1997 Ford Rescue/HAZMAT Van
The CFB Comox, British Colombia, Fire Department acquired this Ford van to carry equipment to support any HAZMAT operations, vehicle extrication work and high-angle rescues. It carries SCBA, HAZMAT protective clothing and a variety of special tools.

1997 International Navistar / Eastway-Paystar 5000 Water Tanker
The role of this water tanker was twofold: nursing other fire apparatus or supplying water directly to hoselines during firefighting operations.

Main features: a 6x6 custom chassis, a six-cylinder 400-hp Cummins diesel engine, capable of cruising at 80 kph / 50 mph, a Hale pump with a capacity of 950 L/min / 210 G/min at a pressure of 1000 kPa / 150 psi, and a 13700 L / 3000 G water tank.

1997 Ford F150 Davtair DA-1000
Main features: a Ford 4X4 chassis, a Ford diesel engine, a 660 L/min / 145 G/min at 140 kPa / 200 psi pump, powered by a Volkswagen diesel engine, a 1000 L / 220 G water tank, a 60 L / 13 G foam tank, a 205 kg / 450 lb PK unit, a bumper turret and hoselines, capable of delivering foam or PK singularly or simultaneously.

2000 E-ONE Cyclone Platform
This vehicle, purchased for CFB Borden, built by E-ONE of Ocala, Florida, was the first of its type to see service in DND.

Main features: an E-ONE custom chassis, a Cummins six-cylinder 4.5 L turbocharged diesel engine, a 30 m / 100 ft elevating platform, and a Hale 5700 L/min / 1250 G/min pump and an 1150 L / 250 G water tank.

2000 E-ONE Aerial
This aerial, purchased for HMC Esquimalt Dockyard, was built by E-One of Ocala, Florida.

Main features: an E-ONE custom chassis, a Cummins, six-cylinder 4.5 L / 275 in^3 turbocharged diesel engine, 30 m / 100 ft aerial ladder, a Hale 5700 L/min / 1250 G/min pump and an 1150 L / 250 G water tank.

2000 Walter P4-6000 MFV
This MFV was built by Walter in Montreal, Quebec.

Main features: a Walter custom chassis, a rear-mounted six-cylinder Cummins diesel engine, producing 586 hp, a 5700 L/min / 1250 G/min pump, a 6000 L / 1320 G water tank, a 360 L / 80 G foam tank, a 225 kg / 500 lb PK unit, a turret, a bumper turret, two handlines and under-truck nozzles.

2001 E-ONE Typhoon Pumper
This pumper was purchased from the E-ONE Company of Ocala, Florida.

Main features: an E-ONE custom chassis, a Cummins 350-hp six-cylinder turbocharged diesel engine, a 3800 L/min / 840 G/min Hale pump, a 3800 L / 840 G water tank, a 115 L /25 G foam tank, a FoamPro foam induction system, with a digital display for water flow-rate and the percent of foam concentrate being inducted and no in-line restrictions.

2002 Western Star Range Truck
Main features: a Western Star chassis, a Cummins 285-hp diesel engine, providing a top speed of 90 kph / 55 mph, 18-hp Briggs & Stratton Wajax pump with a 4500 L / 1000 G water tank, and a 230 L / 50 G foam tank.

Picture Gallery IX

The first fire truck (1920).

1938 Ford crash tender.

1939 GMC/American-LaFrance pumper.

CHAPTER 9 • APPARATUS

1939 International pumper.

1941 International Bickle-Seagrave pumper.

1941 ladder truck.

1941 Code 30 (re-designated G10) pumper.

CHAPTER 9 • APPARATUS

1941 Code 33 (re-designated G15) crash tender.

1942 American-LaFrance pumper.

1942 Class 125 Ford crash truck.

1943 Class 155 Kenworth-LaFrance crash truck.

1944 American-LaFrance aerial.

1949 G17 DC crash vehicle.

1950 Jeep fire chief command vehicle, Goose Bay, Labrador.

1951 MPV range truck.

1951 G18 DC vehicle.

1951 MPV range truck.

1951 G9 triple combination pumper.

1952 Bickle-Seagrave aerial.

1952 Ford crash tender.

1952 Ford pumper.

1952 G11 Bickle-Seagrave pumper.

1952 G11 Thibault pumper.

1952 G13 International DC LRV.

1952 G13 Ford DC LRV.

1953 G11 American-LaFrance pumper.

CHAPTER 9 • APPARATUS

1953 GMC-Bickle-Seagrave pumper.

1953 G21 Thornycroft MFV.

1954 G23 FWD MFV.

1954 GMC-Bickle-Seagrave pumper.

1954 FWD pumper.

1954 Walter Bickle-Seagrave MFV.

1955 International-Thibault pumper.

1956 International Harvester crash truck.

CHAPTER 9 • APPARATUS

1957 G19 Alvis MFV.

1958 Ford range truck.

CHAPTER 9 • APPARATUS

1957 G8 FWD pumper.

1958 E62 street flusher.

1960 Fargo aerial.

1960 Tracked DC crash vehicle.

1962 Ford range truck.

1963 military pattern vehicle pumper.

1964 Thibault pumper.

1964 FWD King-Seagrave aerial.

1964 G19 Sicard MFV.

1965 Ford King-Seagrave pumper.

1965 Walter King-Seagrave MFV.

1966 Thibault pumper.

1970 UNIMOG quick-response vehicle.

1972 International-Pierreville pumper.

1972 Oshkosh M-1000 MFV.

1972 Oshkosh T-1000 MFV.

1974 M.A.N. Gebruder Bachert TCP.

1974 Ford Saskatoon Barton-American pumper.

1975 International King-Seagrave TCP.

1975 Ford aerial.

1976 FLEXTRAC PK crash vehicle.

1977 Kenworth-Thibault aerial.

1978 SCOT Pierre Thibault aerial.

1979 IHC Pierre Thibault TCP.

1979 Oshkosh M-1000 MFV.

1981 International Pierreville aerial.

1981 RIV.

1981 CDN Research MFV.

1982 Ford King-Seagrave TCP.

1982 MPV range truck.

1982 Walter 4500 MFV.

1982 Universal Go-Track crash vehicle.

1984 Western Star range truck.

1984 Ford Thibault TCP.

CHAPTER 9 • APPARATUS

1985 Walter 4500 MFV.

1987 E-ONE Titan MFV.

1988 Ford Superior TCP.

1989 Timberjack range truck

CHAPTER 9 • APPARATUS

1989 Waltek Universal Go-Track crash truck.

1993 Spartan-Thibault TCP.

1993 Oshkosh 6000 MFV.

1995 Navistar-International range truck.

CHAPTER 9 • APPARATUS

1995 Oshkosh T-1000 MFV.

1995 Emergency One P-150 MFV.

1995 Tibotrac water tower/pumper.

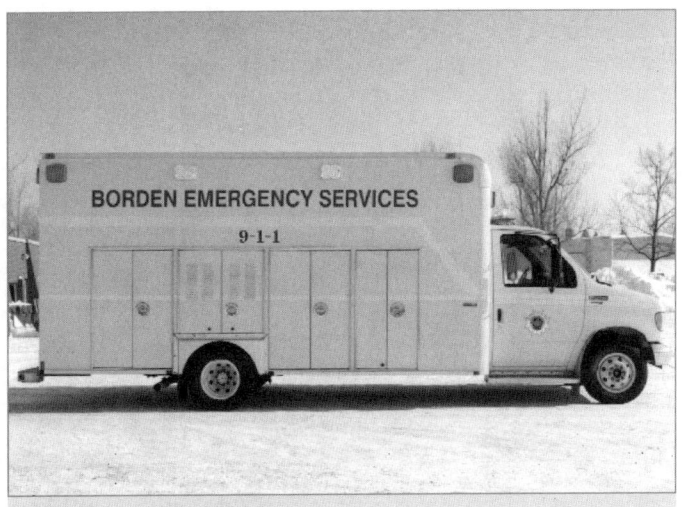
1996 Ford rescue van.

1997 Ford rescue/HAZMAT van.

1997 Ford Davtair DA-1000 crash truck.

1997 Navistar/Eastway-Paystar 5000 water tanker.

2000 E-ONE Cyclone platform.

2000 E-ONE aerial.

2000 Walter P4-6000 MFV.

2001 E-ONE Typhoon TCP.

2002 Western Star range truck.

CHAPTER 9 • APPARATUS

World War II era fire boat, HMCS *Rouille*, Halifax.

World War II era fire tug, No. 1, Halifax.

The fire boat HMCS *Fire Bird*, Halifax, 1975.

The fire boat HMCS *Fire Brand*, Esquimalt, 1975.

CHAPTER X

OPERATIONS

Introduction
In the process of compiling this sampling of Fire Service operations, it was decided to portray the role by describing a number of notable incidents. In doing so, records were perused and the powers of recall of many firefighters were pushed to their limits. Only incidents held to be of special interest for which detailed information could be obtained have been included.

Fire—Shipboard, Halifax Harbour
The Naval Fire Service was involved in several significant incidents, one of the more serious being an explosion in a ship's hold at Bedford Magazine in July 1943. The incident began when the USS *Volunteer*, a freighter loaded with phosphates and magnesium, caught fire while waiting for a convoy escort. The naval fire service from HMC Dockyard Halifax was called to render on-board firefighting assistance. Stoker William Carson had entered a hatch to attack the fire when an explosion occurred and blew him out of the hatch into the harbour. He somehow survived this experience well enough to return to his former job with the Toronto Fire Department.

Fire—Bedford Magazine, Bedford, Nova Scotia[1]
During the summer of 1945, conditions at the Bedford Magazine made it ripe for disaster. With the end of the war in Europe, masses of ammunition were stacked in the magazine from returning warships. Political pressure was applied to unload the ships as rapidly as possible in order to get their reserve crews home, and

---
[1] Archives of HMC Halifax Dockyard Fire Department.

many safety regulations were ignored in the name of expediency. Operating in this manner depends too much on luck, but luck, as events were to show, can become a deserter in a flash. It did just that on July 18, when fire broke out and set off a number of explosions and threatened to blow up the rest of the ammunition. These explosions had the potential for causing great damage to the cities of Halifax and Dartmouth.

At 1835 hours, on July 18, 1945, Petty Officer Emmerson, station officer (platoon chief) with the Bedford Magazine Fire Department (BMFD) of the Navy Fire Service, was notified that there was a fire on the Dauth Jetty at the Bedford Magazine, located towards the northwest end of the Halifax Harbour, and that shells were exploding.

The BMFD responded immediately; however, before they could reach the scene, a major explosion rocked the countryside. This occurred at approximately 1839 hours, presumably involving ammunition on the Dauth Jetty and lighters docked there. The debris scattered by the force of the explosion started numerous natural-cover fires in the surrounding area, some of which were so extensive that they were still being fought the following day.

The Bedford fire crew went immediately to the Dauth area and initially used hoselines from two standpipes before connecting a 2.5 in / 65 mm hoseline from a hydrant to fight fire in grass surrounding the site of some depth charges on the brow of a hill near a detonator storage shack. At approximately 1905 hours, the South Ammunition Depot Fire Department (SADFD) arrived and set up operations in the lower level south area.

Fires were extinguished wherever possible even though major explosions from time to time forced firefighters to retreat. At approximately 2035 hours, the fire crews retreated up the highway toward the town of Bedford, where they were ordered to standby by Navy Captain Robertson and be prepared to re-enter the area.

At 2210 hours, both fire crews were ordered to proceed to the town of Dartmouth and were placed under the command of the Dartmouth fire chief in case of a major explosion, which was contemplated. In the face of this threat, 125000 residents of Dartmouth and Halifax were evacuated.

The firefighting teams fought the tenacious fire, facing exploding shells and bullets that frequently drove them to the ground. After almost two nerve-racking days, the firefighting teams brought the fire under control, and all personnel were allowed to return to their homes. It was not until 1820 hours on July 19 that the BMFD was able to return to a normal state of standby readiness in the Bedford Magazine Fire Hall. It was a demanding test under extreme pressure for the Navy Fire Service.

FR7 Gordon Morrison, later fire chief of HMC Esquimalt Dockyard for several years.

Due to their outstanding performance during this extremely hazardous incident, Lieutenant William Carson, Substantive (SB) RCNVR, Command and Base Fire Chief, Atlantic Command, and Acting Leading Stoker, Firefighter (FF) RCNVR Gordon Morrison were awarded the BEM for bravery. Harold Coxon, Commissioned

Technical Officer (SB) Assistant Naval Fire Chief, Halifax, received a commendation for his actions.[2]

Crash—York Freighter, Edmonton, Alberta[3]

On Thursday, May 26, 1955, at approximately 1420 hours, a four-engine York transport plane was taking off from the Edmonton municipal airport when it hit the instrument landing system shed at the end of the north-south runway, before plowing into a railway yard. Observers of the crash said the aircraft hit the first row of box cars and then seemed to slide over the other rows of box cars before coming to rest on the ground against a tank car and more box cars. One engine landed inside a box car attached to a tanker car, and tail section landed 300 ft / 90 m further north. It was the largest piece left after the fire was out, some two and a half hours later. The Avro York aircraft, a freighter version of the famed Lancaster bomber, was carrying seven tons of goods, mostly food; however, about 100 lb / 45 kg of blasting powder was also on board. The pilot and co-pilot both died in the crash.

The Air Force fire department, which provided fire protection for the airport at the time, was housed in a fire hall attached to a hangar facing the airfield. Most of the on-duty firefighters, plus a number of unmarried firefighters who lived in the fire hall, witnessed the impending crash and were on the vehicles when the crash bell sounded. A G10 pumper and a G13 DC truck responded. The G10 proceeded to the end of the runway where a ditch prevented further progress. Hoselines were laid by hand across the ditch and between the rail cars to reach the fire. However, shortly after these lines were laid, the rail yard engines started removing cars away from the fire area which cut the hoselines, forcing the firefighters to dig the new lines under the tracks. The G13 was dispatched off the base and was able to access the site by driving across the tracks. The City of Edmonton Fire Department responded at 1430 hours and had to lay hose about 2000 ft / 610 m from a hydrant to the north of the rail yards; they connected the line into a pumper and ran auxiliary lines to fire. In all, they dispatched four pumpers, a ladder truck, a hose tender and a squad car.

Once the fire was under control, the search for the occupants began. This turned out to be far more difficult than fighting the fire, in that the wreckage was spread over a huge area, covered with mountains of debris. The search went on for hours until the RCMP and medical authorities were confident that positive identification could be made.

It is interesting to note that in 1955, there were no programs to help people deal with the trauma associated with gruesome incidents of this nature. However, the mess hall had remained open after normal hours and offered hot meals for all. Few could accept the generous offer.

Rescue—Bristol Freighter, Marville, France

No. 1 Fighter Wing at Marville, France, was the site of two disastrous crashes involving Bristol Freighter aircraft. These aircraft were widely used to ferry cargo and passengers during the early years of No. 1 Air Division activities in Europe. The crashes accounted for considerable loss of life, but one in particular had an interesting footnote. The Air Division Fire Marshal, Flight Lieutenant Archie Graham, had the misfortune, on December 1, 1955, to be a passenger on the first one to crash. The aircraft flew into a heavily treed mountain in heavy fog. Fight Lieutenant Graham was lucky to survive the accident, but did sustain a broken ankle and minor abrasions. He scrambled out of the wreckage and began assisting other survivors to drag the dead and injured from the ensuing fire. Later, he hobbled out of

---

2   Memoirs of Lieutenant Commander G.L. Lay (unpublished, May 1993).
3   Material provided by the *Edmonton Journal*.

the woods to a French village several kilometres from the crash site, where he was able to report the location of the downed aircraft.

Fire—Aircraft Hangar, Langar, England[4]
During the 1955–58 time period, 30 Air Materiel Base at the Langar Airport, Nottinghamshire, had dual tenancy: the RCAF and the A.V. Roe aircraft company. The RCAF provided all services, including airfield maintenance, air traffic control, navigational aids and ARFF services, as well as structural firefighting services. In the main, military facilities were located on one side of the airport and the civilian facilities on the opposite side of the airport, where A.V. Roe was conducting experimental work and test-flying the Vulcan bomber and the more aged Shakelton bombers that were being modified for use in maritime patrol. Each of the Shakelton's four engines had unique counter-reciprocating propellers. The Vulcan test aircraft were painted all white, which combined with its revolutionary delta wing, caused them to be quite an awe-inspiring sight at the time. This was particularly so when the aircraft made its final approach on landing, which normally was accomplished with a pronounced nose-high attitude.

On December 17, 1955, a major fire occurred on the civilian side of the airport which brought considerable recognition to the Fire Service. The fire was in an aircraft maintenance hangar that held many 45 G / 205 L drums and other-sized containers of paint solvents, thinners and the like, many of which exploded, as well as Shakelton bombers. The Air Force fire department was the first-in department and remained the sole department involved for an extended period. Although there was extensive damage that included one of the Shakelton aircraft, other aircraft, miscellaneous contents and the structure were saved.

The Nottinghamshire authorities regarded the operation as highly successful and the Nottinghamshire County Fire Service presented the RCAF fire department with a Certificate of Meritorious Conduct. Each member of the crew that was involved in the firefighting operation received a signed copy of the certificate.

The duty crew.
Back Row: LAC MacLean, LAC Greer, Cpl Prawdzik, LAC Maranda and LAC McCartney.
Front Row: LAC Roth, LAC Swain, Mr. McIntyre, LAC Stanley and LAC Murphy.

Fire—Montmendy, France
This incident occurred on July 10, 1956, when a fire started as a shipment of aviation fuel was being unloaded at Montmendy, France, a NATO fuel dump near No. 1 Fighter Wing. Sergeant D. Stevenson, a 30-year-old RCAF firefighter from Montreal, for his role in quelling it, was honoured with the George Medal one of the Commonwealth's highest awards bestowed in peace time for gallantry. The citation for the award of the George Medal reads as follows:

---

[4] Memoirs of Lieutenant Colonel Lorne MacLean (1951–87).

On 10 July 1956, a fire occurred in a shipment of aviation fuel being unloaded at Montmendy, France. Sergeant Stevenson, who was acting as Deputy Fire Chief, performed two acts of bravery that enabled the fire section to bring the fire under control and finally extinguish it. On the first occasion, he climbed to the top of a sealed tank car that was in danger of exploding because of extreme heat and, after repeated efforts, managed to release the manhole cover lock. This relieved the pressure building up inside the tank car and prevented an explosion that would have made this fire completely uncontrollable. Later, while the fire was being kept under

The mess at Montmendy, France.

control but could not be extinguished, Sergeant Stevenson, protected by only a heat mask and a stream of water sprayed on him by other firefighters, again proceeded to the top of the blazing fuel tender and successfully introduced a foam line through the manhole cover, and thus brought the fire under control. Both these acts of bravery were carried out with the full knowledge of their necessity and of the dangers involved. The courage and unselfishness displayed by Sergeant Stevenson and the complete disregard for his own personal safety on this occasion was in the highest traditions of the Royal Canadian Air Force.

Queen Elizabeth presented the medal at Government House, Ottawa, on Dominion Day, 1959.[5]

Crash—AVRO Arrow, Toronto
At 0951 hours on March 25, 1958, AVRO Arrow number 25201 took off from Malton Airport (later renamed Lester B. Pearson International Airport) in Toronto with AVRO's chief experimental test pilot, Jan Zurakowski, at the controls. This maiden flight lasted 35 minutes. The Arrow remains the only supersonic aircraft ever developed in Canada. To put this achievement in perspective, the majority of passenger aircraft using the airport were four-engine piston and turboprop types, such as the Constellation, the Britannia and the Viscount, which would not seem large by today's standards, but were certainly considered heavyweights at the time.

Provision of aircraft rescue and firefighting services for the Arrow project was the responsibility of the A.V. Roe Aircraft Plant Fire Department, which had been supplied with a MFV and a modified jeep by the Department of Transport for this purpose. At about this time, the minister of national defence, George R. Pearkes, VC, decided that the crash protection for the Arrow Flight Test Program was to be beefed up by the temporary addition of an RCAF MFV and crew. Consequently, a short time later, a G19 Alvis MFV from the Fire School at Camp Borden and 13 firefighters from various RCAF stations across the country arrived at No. 1 Supply Depot, Downsview, Ontario.

The organization was set up with Warrant Officer 2 Bill Bourne as chief and Sergeants Brown and Stevenson as crew chiefs. The remainder of the team consisted of two corporals and eight aircraftsmen, resulting in two shifts of six firefighters each to provide protection for all flights of the Arrow aircraft. They were all billeted at RCAF Station Downsview (Toronto) and travelled to and from Malton Airport each day that the Arrow was to be flown.

The A.V. Roe Fire Department kept their MFV, but turned the jeep over to the RCAF firefighters. This jeep had been fitted with two 50 lb CO_2 cylinders and a hose reel, as well as tools, etc. It also had a ground-control radio, which allowed the Air Force crew to "shepherd" the Air Force G19 whenever it was on the airfield, since it was not equipped to receive those radio frequencies.

Whenever the Arrow was scheduled to fly, the RCAF crew would suit up and move the MFV and jeep out to the runway area for the take-off, and then wait for its return. Once it was back down and had returned to the A.V. Roe tarmac, if no more flights were planned for that day, the crew would be through. As crash-duty goes, it was a pretty good deal.

On Wednesday June 11, 1958, Arrow No. 201 took off just after 1400 hours on its eleventh flight, which

---

[5] *Firefighting in Canada*, August 1959.

would last about an hour and a half. Sergeant Stevenson as the on-duty crew chief was the jeep driver and positioned it behind the A.V. Roe "follow-me" vehicle at a point about mid-way down runway No. 32, about 200 to 300 ft / 60 to 90 m off the eastern edge. The G19 and crew were located at the far end of the runway in company with the A.V. Roe MFV. From the latter position, the runway button (threshold) was over a slight rise and out of sight.

On this fateful afternoon the aircraft rescue and firefighting crew watched as Arrow No. 201 made its approach to the runway until it dropped out of sight and before rolling up over the rise to pass in front of them. They continued to watch its progress until it disappeared behind the "follow-me" vehicle, which was now obstructing the view. Due to this, Sergeant Stevenson leaned over to his right to pick the Arrow up again and after a moment began to suspect that something was wrong since it still had not reappeared from behind the obstruction. He then pulled the jeep to the right around "follow me," and there was No. 201 sliding sideways with its tail towards the oncoming jeep, off the far side of the runway, squirting dirt straight up into the air.

The aircraft came to rest almost directly in front of the "standby" position where the ARFF crew was located,; a record may have been set for the shortest "time of arrival" at a crash site. The Arrow had lost its undercarriage and was down on its belly.

Flying Officer Stevenson recalls that:[6]

> The right wing tip was about 18 inches off the ground so I was able to jump up onto the wing on the dead run and continue on to the cockpit. The "clam shell" canopy had been popped open and pilot Jan Zurakowski "Zura" was standing up facing backwards and I believe I startled him when I yelled, "Are you alright, sir?"
>
> He was unhurt, so after inserting the safety pins into the Martin-Baker ejection seat, I held his helmet for him while he climbed out. Someone told me years later that they had heard "Zura" talking about the incident in an Air Force officers' mess somewhere, and that he said something like: "I don't know where that Air Force Crash Crew came from; they just appeared out of nowhere!"

This accident was caused by a malfunction of the left-side landing gear. The pilot had no indication that a problem existed until after the aircraft had touched down and he found the steady swerve to the left uncontrollable. Yet, upon arrival at the aircraft, thought to be not more than 20 seconds after it came to rest, there was not even a whisper coming from the engines. This extraordinary pilot had shut everything down long before he stopped sliding. The crash crew had to be very conscious of running engines on the Arrow aircraft, because the very large intakes were located just aft of the clam-shell canopy and the Emergency Canopy Release.

Thoughtful Experiment
An interesting experiment was tried at No. 3 Wing, Zweibrucken, in 1963. Inspired by Senior Air Traffic Control Officer Squadron Leader Ernie MacLaren and Fire Chief Flight Lieutenant Johnny Cowell, the station photo section mounted a movie camera on top of an MFV. The idea was to film an incident and surrounding activity when the fire crew responded to an airfield emergency. In this manner aircraft accident investigation teams would be able to review the movie at their convenience.

---
6   Memoirs of Flying Officer Doug Stevenson (1946–63).

After overcoming some initial teething problems, the experiment proved successful. It supplied the necessary photographic evidence to exonerate a pilot accused of an inappropriate emergency landing procedure. The pilot faced the real possibility of being court marshalled until a review of the movie taken from the top of the crash vehicle clearly showed the accused pilot was not in error. Notwithstanding its immediate success, the harsh environment for the type of camera used posed difficulties and was a key factor in discontinuing the practice, although it was still being used in the late 1960s to early 1970s in Lahr.

Crash—CF101 Voodoo, Chatham
The following is from the memoirs of Captain Bob Pineault, as recalled in 2003:

> To set the scene it is helpful to mention that the Chatham area of New Brunswick receives a lot of snow during most winters. Consequently, the snowplows and snowblowers were kept quite busy in order to keep the runways open 24 hours a day. As a winter progressed, there was usually much packed snow along each side of the runways. As well, snowblowers were used to keep an alleyway cleared at least one machine width beyond the runway parallel-lights. As a result, it was not uncommon to have a hard packed snow bank cut perfectly square of least 7 to 8 ft / 2 m to 2.5 m in height bordering the runways.
>
> On this particular day, which I recall as being during the winter of 1966, the infamous "one-bell"[7] crash alarm was sounded by the air traffic control tower. The reason for the alarm was that a CF101 Voodoo aircraft had apparently experienced a failure of its landing gear and had veered off the runway and plowed through the big snow bank and to come to rest upside down approximately 50 ft / 15 m off the runway.
>
> On that day, being a junior firefighter, I was turret man on a G19 Sicard crash truck, which was normal assignment, since only the more experienced firefighters were deployed to operate the crash trucks. Accordingly, as a junior firefighter, you were usually assigned as turret-man, rescue-man or a handline-man. The beauty being turret-man is that it placed you at least 10 ft / 3 m off the ground, which was a favoured observation point from which to view an incident such as this.
>
> As we arrived at the accident site, a little plume of smoke was coming from the aircraft, and I immediately laid a blanket of foam over the area. As I watched, other crew members were busily digging out the pilot and navigator, who were stuck in the cockpit of the upside down aircraft. I remember that we had to dig them out with shovels. Once freed, they hastily departed the area.
>
> As will be seen, despite the graveness of the situation, it was not without a touch of humour, in this case, at the expense of Fire Chief Flight Lieutenant Bill Muirhead. The Chief, of exceptional gentlemanly character, was also invariably exceptionally well turned out in uniform as he was on this occasion. As this accident had unfolded, the aircraft landing gear had cut deep furrows in the high bank of snow, which had now been

---

[7] A "one-bell" crash alarm was notification that a crash had occurred, whereas a "two-bell" alarm was notification of an aircraft in-flight reporting a problem that was a threat to its safety.

filled with foam to the extent that they were no longer visible. As luck would have it, as Bill moved toward the stricken aircraft, he stepped into one of these hidden trenches and momentarily disappeared from sight. When he reappeared and climbed out of the hole, he brushed the foam from his uniform, calmly put his hat back on and continued to the aircraft as if nothing happened. After all, Flight Lieutenant Muirhead had an array of medals on his uniform that had been earned as an air-gunner during WW II and something like becoming enveloped in foam did not faze him in the least.

Crash—Air Canada DC 8, Ottawa[8]

The following is from the memoirs of Chief Warrant Officer Jim Lockhart (1951–84) as recalled in 2004:

> We had commenced our night shift at 1800 hours on May 19, 1967, and I, Sergeant Jim Lockhart as platoon chief, had just finished detailing the firefighters to their positions on the duty board prior to entering the Control Room when the crash alarm sounded. As I looked out the window towards the airfield I could see a large tower of smoke in an area of the airfield that was some distance from the fire hall. If memory serves well, there were probably six of us on duty that evening. I immediately ordered a recall of off-duty firefighters and had the E62 flusher put on standby.
>
> As we rolled out of the fire hall, I contacted the air-traffic control tower for a description of the problem and was informed that an Air Canada DC 8 arriving from Montreal with 87 or 89 souls on board had crashed, and was given the location. We had no problem seeing the general area of the crash, which was vividly marked by a tall plume of smoke. While travelling towards the scene I radioed the fire department watch-keeper for a SITREP on the recall. He informed me that it was progressing well and that he had already contacted some of the firefighters who lived on the base and they were on their way in. I then instructed him to bring the flusher vehicle to the crash site.
>
> The initial response included one LRV and two MFVs. The LRV and one MFV were the first on the scene, and the fire was quickly knocked down with the exception of some "ghost-flames"[9] here and there. At this point we ran into a snag when the second MFV, which was situated approximately 30 to 40 yds / 27 to 37 m from the downed aircraft, failed to respond to a signal to come forward. I directed one of the firefighters from the first-in response and to bring it forward to the immediate fire ground area to control any flare-up of fire that might occur. Upon arrival at the vehicle, the firefighter found that the operator had been incapacitated by the sudden onset of an illness. He assisted the stricken operator from the vehicle and repositioned it as required.
>
> Coincident with the foregoing, firefighters were searching the area around the site, looking for casualties. They reported that

---

8 A Web site maintained by Air Canada, www.jacdec.de/Air%20Canada.htm, contains this terse record: "19.05.1967 Douglas DC 8–54F CF-TJM 45653. On approach to Ottawa AP (at18:37L) the aircraft rolled to the right and struck the ground in an inverted position. The three crew members died."
9 See Appendix "B" for definition.

they could find none. I contacted the tower to confirm the information I had received was correct. I was then informed there was a correction and that it was a training flight from Montreal with three persons on board, two pilots and a third crew member. After receiving this updated information, there was no problem in locating the crew, removing them from the crash, and placing them in the ambulance.

The second MFV extinguished any remaining fires as part of the overhaul process. The E62 and pumper were used to cool the area down and recharge the first-in MFV. This was done by firefighters who had responded to the recall. When all of this was completed the accident site was turned over to other airport authorities and we returned to the fire hall and critiqued the overall operation.

The final chapter to this incident occurred some time later when the fire chief, Captain John Cowell, attended a dinner sponsored by Air Canada to recognize some of the participants that responded or assisted at the crash. The airport manager congratulated Captain Cowell on the efficiency of his department in handling the emergency and on receiving a medal. Captain Cowell, I believe, then informed him that it was not our department that received the medal and that it was some person or organization in the medical field that was the recipient.

Fire—Motor Transport Building, Ottawa
At 0300 hours on December 6, 1967, a minor explosion occurred in the base transportation building, activating the sprinkler system which, in turn, caused an alarm to be transmitted to the fire department almost directly across the street. The responding crew, led by Sergeant Pat Finnigan, entered the building to find a man-door to the compressor/sprinkler room blown off its hinges. Entry was made by Corporals Yvon Auger and John Gordon in CHEMOX SCBA with direction to shut down the sprinkler system, as no fire was visible. Privates Peter Boyce and Brian Lobb were providing coverage for the entry team from the doorway. As the entry team reached the Outside Screw and Yoke (OS&Y) valve, all hell broke loose. There was a major explosion centred in the compressor/sprinkler room, which knocked down a full cement block wall between the compressor/sprinkler room and the tool crib, as well as portions of the other interior cement block walls. The roof of the transport section was shifted 10 in / 25 cm, and the huge sliding fire door between the vehicle maintenance and storage bays was knocked to the floor.

A leaking "Bowser" aircraft refuelling vehicle in the maintenance bay apparently overburdened a poorly maintained fuel/water separator located directly beneath the compressor/sprinkler room. Vapours from the leaked fuel that were ignited by the start-up of the compressor caused in the initial relatively minor explosion. As more fuel escaped, it continued to rise on the surface of the water from the sprinkler system to a level that, when the automatic compressor started, it resulted in a major explosion. The ensuing fire was battled from the floor area and upon the roof, using both pumpers and foam vehicles to achieve extinguishment.

In retrospect, it is astounding that no lives were lost during this violent event, even though six of the nine firefighters who responded were injured. Two were hospitalized, Private Boyce with facial burns and Private Lobb with head injuries. Deservedly, the fire crew received several awards in recognition of their outstanding performance. These are listed below, followed by a certified true copy of the letter of commendation from the Commander, Air Transport Command.

> Queen's Commendation for Bravery:
> Sergeant Pat Finnigan
>
> Commander Air Transport Command Commendation:
> Corporal Yvon Auger
> Corporal John Gordon
> Private Peter Boyce
> Private Brian Lobb
>
> Gold Helmet award by a private company:
> Private Peter Boyce, and
> Private Brian Lobb.

Rescue—Cold Lake, Alberta[10]

The Air Force magazine *Flight Comment* published the following article:

> On the night of September 18, 1973, in Cold Lake two T33 aircraft were reported missing. To conduct the initial search, a helicopter with a pilot, a helicopter crewman and firefighters Master Corporal Clackson and Corporal Neufeld was launched.
>
> Shortly after takeoff the helicopter picked up an emergency homing signal and soon arrived over one of the downed pilots who had a small fire going. Because of the high trees, the pilot was unable to land in the area. The helicopter crew chief therefore asked if the firefighters if they wished to volunteer to be lowered in the sling hoist. This was Corporal Neufeld's first flight in a helicopter and although he had never used a sling hoist, he immediately volunteered.
>
> Corporal Neufeld located the downed pilot, who was in a dazed condition, and walked and assisted him to the sling. He then secured him in the sling and signalled for him to be hoisted to the aircraft. Corporal Neufeld then returned and extinguished the camp fire, recovered the pilot's emergency beacon and was himself hoisted into the helicopter.
>
> The crew continued to receive an emergency locator signal so they proceeded to home in on the signal until another camp fire was spotted. The helicopter was once more was forced to hover because of the terrain, and Corporal Neufeld was again lowered by hoist. The second pilot was located by his camp fire and, because of knee injuries, was assisted into the sling by Corporal Neufeld. Corporal Neufeld again extinguished the camp fire, retrieved the pilot's personal locator beacon and was then hoisted aboard.
>
> The quick rescue of these two pilots injured and in a state of shock is directly attributable to Corporal Neufeld's courage and devotion to duty well beyond the call of his normal duties as a firefighter. The fact that it was his first flight, over unknown terrain at night, only serves to emphasize this act of rescue. In recognition of this, *Flight Comment* gave him a "Good Show" award.

---

10   *Flight Comment, Good Show Award*

Fire—Fuel Farm, Courtney, British Columbia

On a pleasant Saturday morning, September 28, 1974, the weather was sunny and the winds quite calm as a tanker vehicle was taking on a load through an open hatch from an overhead feed line from a bulk storage tank in the city of Courtney. The horizontal section of the feed line was iron pipe, and the vertical drop to the vehicles was a flexible hose with a shut-off valve at the discharge end.

The fuel farm was located on high ground, somewhat like a knoll, and the adjacent loading dock was at a lower elevation, which provided for the gravitational flow of fuel from the base of the bulk storage tanks to load tanker vehicles. A concrete wall encircled the fuel farm to serve as a dyke. There was an open, grass-covered ditch running northward down a slope along the street parallel to the loading dock. The ditch ended at a flat area sloping slightly in a westerly direction some 200 m / 220 yds north of the fuel transfer area.

The serenity of this scene was soon to changed quite dramatically. A few moments of inattention on the part of the tanker operator caused him to fail to notice that the tanker was overflowing. There was considerable fuel around the vehicle by the time he got the fill line turned off. He then entered the cab, started the engine, and a fire flashed over the surrounding area and the vehicle. Luckily, the driver escaped with some painful, but not life-threatening, burns.

From this point on the situation deteriorated rapidly. The tanker became fully involved; there was a large area of fire around the fuel transfer dock and, in short order, the flexible hose burned off. That was when the real trouble set in. The fire was now being fed by a virtually inexhaustible supply of fuel from iron-pipe feeder line. The flowing fuel also added a three-dimensional fire to the mix.

Working under the provisions of a mutual-aid agreement, the fire departments of both the city of Courtney, under command of Chief Lawrence Burns, and CFB Comox, under command of Master Warrant Officer Lloyd Howlett, responded. The base fire chief, Major Lorne MacLean lived in Courtney and responded directly to the scene. The fire was not difficult to find as the intense heat from the liquid fuel fire drove a plume of black smoke high into the air.

Essentially, the strategy and tactics employed by the joint forces was for the city to apply water to the combustible components of the tanker and to direct hose streams onto the bulk fuel storage tanks, as a means of mitigating the effects of the intense heat and the base fire department to extinguish surface fire by applying foam from a MFV. This was quite successful, especially in extinguishing the fire in a large surface of the fuel transfer area. Foam is, however, ineffectual against a three-dimensional fire, as presented by the stream of flaming fuel descending from the filler line. The vehicle and the immediate surroundings continued to burn fiercely and scorch the bulk storage tanks.

Upon arrival at the scene and assessing the situation, the base fire chief ordered the cessation of foam application on the three-dimensional fire, the application foam only as necessary to maintain the integrity of the foam blanket, and that a supply of foam concentrate be brought to the scene from the base.

The situation at this point was unique in that the stream of burning fuel continued to bathe the tanker vehicle in flames. As the mix of fuel and water moved northward, the flames were being extinguished as the mixture slid under the foam blanket and then emerging as a liquid and thence flowed down the open ditch. Bordering the flat area where the ditch ended was a housing development. The scene for a potential disaster was being set. Somehow the flow of fuel from the feed line had to be stopped or evacuation orders issued. The two fire chiefs decided that an attempt would be made to pass into the intense heat and enter the compound. The base fire chief asked Firefighter Corporal Armstrong if he would accompany him on the undertaking. He agreed to do so. With the city fire crew spraying water on them and the storage tank, they got inside the wall of the dyke only to find that the shut-off valve was chained and padlocked in the open position. This forced the team to return to a fire truck to get a set of bolt cutters and then re-enter the compound. The flow of fuel was stopped and the incident was brought to a successful conclusion.

As a result of CFB Comox Fire Department's actions in fighting this fire, Major Lorne MacLean and Corporal Don Armstrong were awarded the Chief of the Defence Staff's Commendation.

Rescue—McIvor Lake, British Columbia
On June 19, 1976, a near tragic accident occurred at McIvor Lake near Campbell River, British Columbia. A few days later, an article by Bill Mathis under the heading—CFB Firemen Rescue Boy—was published in the local Comox newspaper. It relates the following story:

> It's something like a fireman being honoured for putting out a fire, said CFB Comox Fire Chief, Captain Don Carmichael, in trying to underplay his role in rescuing a 12 year old Campbell River boy.
>
> Carmichael and base firefighter Corporal Ed Neufeld have been nominated to receive an award from the Royal Lifesaving Society for their actions June 19 at McIvor Lake.
>
> Carmichael, a former lifeguard and swimming instructor, disdains the role of hero, saying "I just happened to be out at the lake at the right time." He gives credit to Neufeld, who applied mouth to mouth resuscitation to young Sam Quatell after Carmichael fished him out of the lake.
>
> Neufeld and Carmichael were camped by the lake eating supper when a neighbouring camper asked if someone could swim because a boy was apparently at the bottom of the lake. Family member Lena Quatell said Sam was swimming with his 23 year old brother Don, when he ran into trouble. "They were out in the water when a family friend Ritchie Drake asked Don if Sam could swim. Don Quatell just froze." She said Drake screamed at Don to swim out and get the boy but the brother was unable to move. "I saw Sam's head come up once, go down and come up again and then disappear."
>
> After being alerted to the problem, Carmichael grabbed his son's snorkel and face mask, stripped of his clothes and raced to the water. Carmichael said the water was cold and murky. "As luck would have it I swam over the top of this young fellow, but didn't see him at first," he said. "The water was so churned up it was hard to see, but the mask made the difference." Carmichael passed the boy to Neufeld who began resuscitation while carrying him from the water. Initial efforts to revive him seemed hopeless, since he had turned blue and wasn't breathing at all. Sam had a bowel movement, a common occurrence when a person approaches death. But bystanders noticed body movement and with continued effort Neufeld revived the boy. With Sam

wrapped in blankets, Carmicheal and Neufeld headed their travel trailer for the hospital, but were met by en-route by an ambulance.

Neufeld credited the training in mouth-to-mouth resuscitation which firefighters receive for the successful revival. "I kept remembering to keep the boy's head back."

The boy recovered from the near-drowning without complications.

Capt Don Carmichael and Cpl Ed Neufeld.

Crash—Argus, Summerside, Prince Edward Island[11]

Overview
On Thursday, March 31, 1977, an Argus aircraft, tail number 20737, from No. 415 Squadron, crashed as it was attempting to regain control and altitude after an aborted landing attempt on runway No. 18.

Fifteen of the 16 crew members on board survived the immediate impact of the crash which occurred after the Argus struck a Lockheed Electra aircraft parked on the ramp in front of No. 4 Hangar. Overall, three crew members died as a result of this crash: Major Ross Hawkes, Sergeant Ralph Arsenault, and Master Corporal Al Senez. Although Major Hawkes was the aircraft captain, the co-pilot was flying the aircraft when the landing was attempted. Unfortunately Major Hawkes was standing behind the co-pilot and, therefore, did not have the benefits of the restraints that would have been in place had he been in his seat. Six other crew members suffered many injuries, including burns; however, it is believed that all recovered and returned to duty.

A second plane that became part of the overall accident scene was a Nordair Lockheed L-188 CFNA. It was parked on the ramp and was clipped by one wing tip of the descending Argus.

On the day of the crash, the CFB Summerside Fire Department was equipped with three G19 MFVs, one LRV and two PUMPERs. One of the MFVs was out of service due to a prime mover engine rebuild. The duty platoon was working at the minimum duty strength of seven, inclusive of the fire department dispatcher. Warrant Officer "Chuck" Coolen was in command, and Sergeant Jack Alexander was second-in-command.

Events Leading Up to the Crash
On the day of the crash, the weather was rather cool and blustery. The temperature was near 0°C and the wind was gusting out of the south at up to 30 knots. At 1040 hours, the duty air-traffic controller contacted the fire department and reported that an Argus aircraft was returning to base with Engine No. 1 shutdown due to mechanical problems. Its estimated time of arrival was 1250 hours.

---

11 Memoirs of Sergeant Jack Alexander (1960–88).

News of an Argus aircraft returning to base with an engine out did not generate a lot of excitement. Such occurrences were almost routine events. During lunch, one of the off-duty firefighters dropped by the fire hall and, upon learning that an aircraft was coming in with a problem, volunteered to stay and respond with the crew to meet the aircraft. This meant that manning for the vehicles would be seven, with a driver and turret operator in one MFV, a driver, turret operator and deputy platoon chief in the second MFV, and a driver and platoon chief on the LRV.

At approximately 1245 hours, the crash alarm in the fire hall sounded twice, which indicated that an aircraft with an airborne emergency was nearing the airfield. The two MFVs and the LRV responded and upon gaining radio contact with the control tower were informed that the Argus was on approach for a landing on Runway No.18 with 16 souls on board and 23000 lb / 1040 kg of fuel remaining. The platoon chief directed his vehicles to take up the normal, predetermined standby positions where the taxiway intersects with the runway. This location was about two-thirds of the way down the runway and provided a clear view in all directions.

While proceeding along the tower ramp to the standby positions, it was noted that a Lockheed Electra, which operated on ice patrols out of Summerside each spring, had returned to base and was being serviced by a crew at its normal parking location in front of No. 4 Hangar. As the crash trucks were arriving at the standby location to await the Argus, steady rain began to fall, the wind started gusting strongly, and several claps of thunder were heard.

At approximately 1300 hours, the Argus came into view and appeared to be making a normal approach to the runway. However, during the last couple of seconds of descent before touching down, the aircraft suddenly veered to the starboard and the left undercarriage touched down approximately 100 ft / 30 m to the left of the runway. The aircraft went to full power and began to stagger back into the air. Due to its No. 1 engine being out, full power on the other three engines resulted in the aircraft drifting to port and it was coming directly at the crash trucks at an estimated altitude of approximately 10 ft / 3 m. Immediately, the crash truck drivers manoeuvred to get out of its incoming flight path. As the aircraft advanced, it gradually gained a little more altitude and suddenly started to roll to port. By this time, in just a matter of a few seconds, the aircraft was continuing to veer to port and was heading directly towards No. 2 Hangar, beyond which, on a slight hill, was the base recreation centre, barrack blocks and approximately 150 m / 165 yds beyond them were the married quarters.

Just as it appeared that the Argus would crash into No. 2 Hangar and the recreation centre, it rolled left with its wings almost perpendicular to the ground and began to turn to the left (like an elevator turn, since its wings were still vertical in relation to the ground). The aircraft was now still drifting towards the hangars, but was beginning to move along the taxiway in an easterly direction with the tip of the port wing only a couple of metres above the tarmac. For a moment or two, it looked like the pilot could make it if he could roll the aircraft back level to the ground. The aircraft was heading in an easterly direction towards Malpeque Bay. Suddenly, the left wing tip of the Argus sliced through the fuselage of the Electra on the tarmac about halfway between the wing and the tail. A split-second later, the Argus nosed downwards, and the left wing of the Argus hit the tarmac followed by an explosion and fire ball.

Fire Ground Operations
At this point, the crash trucks were already rolling down the taxiway chasing the Argus. As the initial explosion and fireball dissipated, all that became

visible was a huge fire-front running for an unseen distance easterly along the taxiway. The platoon chief contacted the fire dispatcher to initiate an emergency recall and the control tower called base operations, requesting that all available medical personnel respond immediately to the scene. The fire area was so great that no portion of the Argus could be seen. The port side of the Electra was engulfed in flames, and the fire was spreading across the tarmac towards the hangars. The status of the crew that had been servicing the Electra was unknown. Firefighting operations were started at this location and continued along the front edge of the fire area and eventually located the Argus wreckage about 330 yds / 300 m away.

The initial fire attack by the MFVs immediately ran into problems. Despite full throttle to the pump engine, the resulting reach of stream from the turret was only about 20 m / 65 ft. The pump engine would just not achieve the level of revolutions per minute needed for good pump performance. The second MFV pump engine failed to throttle up at all. In an exceptional display of knowledge and quick thinking under stress, the driver of the truck climbed out, opened the engine compartment and found that the throttle linkage to the pump engine had separated (due to metal fatigue it was found later). He was able to get the pump working by operating the carburetor linkage by hand and his turret operator was able to apply foam effectively to the fire. Simultaneously, one of the off-duty firefighters who had dropped by the fire hall, arrived on scene with the primary pumper. It entered the fray by applying AFFF onto the fire around the Electra by using a 1½ in / 38 mm hose-line connected to a foam inductor to help bring it under control. About this time, the first MFV was performing better and foam was applied to knock down the rest of the fire in the immediate area. About one minute into this firefighting operation, a violent thunderstorm, accompanied by heavy rain and a wind shift of 180 degrees with gusts to 58 mph / 93 kph, drove the smoke and heat back on to the firefighters. Fortunately, this only lasted for three or four minutes before the wind shifted back out of the south.

Going back a bit to just before the Argus hit the Electra, the E62 Flusher, which responded from base transport to take up a standby position at the base of the control tower beside No. 3 Hangar, had been abandoned by its civilian driver when he saw the Argus headed directly towards him. Later on when the MFV, being operated by hand-controlled throttle, ran out of water the operator, noticing that the E62 was only about 65 yd / 60 m away, climbed out of the engine compartment, ran over and drove it over behind the two MFVs, made the necessary hose connections, and began supplying water to the MFVs so they could continue to apply foam. About this time, the fire around the Electra was pretty much knocked down, except for that being supplied by fuel running out of the trailing edge of the port wing which had been ripped open by the Argus.

Coincident with the foregoing, the platoon chief, having arrived at the Argus wreckage, and having expended the 1000 lb / 450 kg of PK carried on that vehicle, ordered one of the MFVs to his area to apply foam to the area around the Argus in an attempt to isolate it from the main body of fire. Having topped up his water tank from the E62, the driver of the second MFV proceeded there to assist as he could. Meanwhile, the second pumper, manned by off-duty firefighters, had arrived at the Argus and was applying AFFF with handlines.

Back at the Electra, the flowing fuel fire from the wing could not be extinguished with a combined attack of three 30 lb / 14 kg DC extinguishers, so a crew member was sent to get some help from bystanders and get the 300 lb / 135 kg DC unit from No. 3 Hangar. Unable to obtain assistance from the onlookers, he pulled the unit out of the hangar by himself. This unit was used successfully to extinguish all remaining fire in the vicinity of the Electra. A search was carried out for any victims around the Electra, and thankfully, none was found. At this point, hangar-line personnel were posted with DC extinguishers and a 50 lb / 23 kg CO^2 wheeled unit as a fire guard, and the three remaining firefighters and the MFV proceeded down to the Argus.

At the Argus aircraft, after it came to rest, almost all of the crew members, even those injured, had managed to extract themselves by one means or another. As they exited the wreckage, they were assisted and comforted by trainees from the Marine and Fisheries Training Centre in Summerside. This group, led by Marine Services Officer (not Navy) John Burke, had been conducting firefighting training in our training area which was only about 90 m /100 yds from where the Argus came to rest. Captain Burke and his trainees unselfishly pitched in without being asked. They not only comforted and aided the survivors, but extinguished small spot fires in the area. The fact that these people had only received five days of shipboard-type firefighting training, as part of their pre-sea duty training prior to being employed aboard ships, makes their actions all the more remarkable and reflects a great deal of credit on Captain John Burke and his training program.

As things progressed, the fire crew learned that one aircraft crew member was still unaccounted for. This member, who had suffered fatal injuries, was finally located partially buried under the front fuselage wreckage that had piled up in front of the starboard wing which was still attached to the remainder of the fuselage. There was still some minor fire in one of the engine nacelles from burning magnesium that could not be accessed due to wreckage on top of it. At approximately 1515 hours, the initial response crew was stood down and returned to the fire hall.

Sadly, two of the survivors died later that day in hospital from injuries resulting from the crash.

Indelible Personal Memories—One
From the memoirs of Sergeant Jack Alexander (1960–88) as recalled in 2004:

> Even still today, 27 years later, I occasionally replay these events back in my mind. I can still vividly see the left wing tip suddenly sliced through the fuselage of the Electra on the tarmac about halfway between the wing and the tail. A split-second later, the Argus nosed downwards and the left wing of the Argus hit the tarmac followed by an explosion and fire ball.
>
> Despite the unexpected event of two large aircraft, a thousand feet apart, totally involved at the same time, the initial minimal manpower, the vehicle deficiencies, the weather, and the other unknowns that occurred, I cannot speak too highly of the manner in which my fellow firefighters and emergency teams, such as the military police and medical staff, responded. The effectiveness of the recall of off-duty firefighters was amazing. After the initial few minutes, I recall that every time things began looking critical, I only had to turn around to find more and more off-duty firefighters arriving on the scene and pitching in.
>
> Similarly, seeing one of my young privates running down the taxiway, pulling that 300 lb / 135 kg DC wheeled-unit fire extinguisher, without help, still amazes me. If he had tripped and fallen, chances are that extinguisher would have rolled right over him and killed him.

Then there was the MFV operator who had the presence of mind to climb down into the pump-engine compartment and control the throttle by hand and, later, had the initiative and knowledge to procure the abandoned E62 and supply us with much needed water. The nearest working fire hydrant was about 500 ft / 150 m from us, and we did not have sufficient manpower at that time to lay a hoseline that distance.

The Marine and Fisheries Training Centre trainees' actions were superb and helped relieve the burden of worrying about taking care of the survivors initially. Captain Burke, an associate and friend at the time, will always be remembered for his cool and calm disposition during the entire event. And last but not least, Master Warrant Officer Art Lamperd, the deputy fire chief. (The base fire chief, Captain Jim Wright, was away on temporary duty on the day of the crash.) As the acting fire chief, Art was like a rock for us. When he arrived on scene, he coordinated our recovery actions with the various base emergency agencies, giving us full support all of the time.

Indelible Personal Memories—Two
From the memoirs of Military Police Captain Carl Delaney (1956–92) as recalled in 2004:

YES! I was there on that terrible day. I remember it well.

I lived in the married quarters area of the base and walked home from the guardhouse for lunch. I walked back at about 1300 hours on that March day and the walk was one to remember, as it was a terrible day in PEI. It was raining heavily with very high winds with gusts over 100 kph / 60 mph which disrupted our telephone system and electricity services, in part due to flooding in the area where all our phone systems wiring was located. As I entered the Guardhouse, the in-flight emergency alarm sounded as an Argus which was on a mission to the Newfoundland area experienced one engine out and was returning to Summerside. As the MWO in charge of the Military Police (MP), it was not normal for me to respond; however, since I was already wet, I responded in lieu of someone else, in company with fellow MP Duke Snyder.

At the foot of the tower we awaited the arrival of the Argus that had Engine No. 1 shut-down—a situation that not unusual or particularly alarming. As the Argus approached, all hell broke out. At this point, the cross winds were very high, and although the runway was clear of snow, there were high banks on both sides. The captain was not in control and this was a check-out flight. Unfortunately, there was a moment of indecision when the captain ordered the aircraft to be turned over to him. The co-pilot hesitated and did not relieve himself of control. Moments later a guest of wind put the Argus off course and as a result the propeller of Engine No. 4 hit the bank of snow, rendering it unserviceable. Only two of the four engines were now operating.

The decision at the time was to avoid a crash, and the co-pilot tried to get as much lift as possible. Unfortunately, after the second engine went out, the aircraft was pointed downwind, which would not allow much lift. The Argus made a turn to the north just missing the tower with the left wing pointing down. Shortly thereafter, the left wing of the Argus cut an aircraft that had been just refuelled and was on the tarmac strip and getting ready for an ice patrol in the North. After the wing sliced the top of this plane, a part of the Argus wing fell off and this caused the Argus to level itself and proceed down the taxi-strip just in front of the hangars. The collision caused a fireball, and fuel was exploding as the Argus was in its fatal descent.

The Argus hit the ground and left the hard surface onto the grass and then it hit the road leading from the fire hall to the runways. As it hit this road it broke open about the centre, which allowed some of the crew to get out. As I recall, those in the aircraft who were facing the tail saw the crack and dove to the floor and out the crack to safety. Those who were facing the front were not so lucky, and the fire ball exploding behind them caught up to the Argus. Engine No. 1 broke free on impact and cut right through the plane to the side closest to the fire hall.

On this fateful day, many factors beyond anyone's control made the tragic event more complicated, including the fact that all the senior members of the Base Hospital staff were in Charlottetown on a training seminar and the Base Medical Services was at a critical manning level, and that the violent weather interrupted both electrical services and telephone services. Back at the Guardhouse, to cover in part for the loss of telephone services, the town of Summerside police chief and I used his police car parked outside my window to relay emergency responses to the local hospital, the Charlottetown burn centre and to the RCMP, who provided escort services for vehicles in conditions of virtual zero visibility.

Finally, I believe we sent a crew to both hospitals and especially to Charlottetown.

The aftermath.

Fire—Use of MXF on a Building Fire, Lahr, West Germany[12]

Sometime in early 1982 a series of fires occurred that was of special interest. At that point the fire department at CFB Lahr had been training with MXF application in respect to structural fires for more than a decade. This series of fires, that were of a suspicious nature, occurred in the downtown MQ, although at the time there was no conclusive evidence that they were the

---

12 Memoirs of Captain Ev Evans (1961–93).

work of a serial arsonist. The outbreak included two major fires, the first in the basement of a ten-storey PMQ building located opposite the Banhoff (railway station) and another in a four-storey building located close to the town of Langenwinkel.

Extinguishing the fire in the ten-storey tower was achieved with a mixture of high-volume water streams and, in the later stages of the operation, with MXF. Although the MXF was seen to be very effective, its ability to achieve complete and expeditious extinguishment of a major structural fire on its own merit was left in question.

A week or so later, another basement fire blazed in a four-storey MQ tower; this structure was of concrete construction. The basement which was approximately 3000 ft^2 / 280 m^2 in area, divided into individual storage units for the residents' use. The separation was achieved with the use of chain-link security lock-ups. These units contained all manner of combustible and flammable material, including quantities of highly flammable liquids, such as camping fuel, and troublesome combustibles, such as vehicle tires.

This fire was attacked and completely extinguished with the exclusive use of MXF, making it a highly successful operation. The potential advantages of using MXF on structural fires are as follows:

- reduces water damage to structures and material contents;
- less demand on municipal water supplies;
- rapid suppression of smoke and other toxic fumes;
- facilitates the early return of occupants;
- increases fire investigators ability to determine the cause of the fire;
- quicker suppression of Class B contents;
- reduces time it takes to control difficult fires, involving material such as vehicle tires; and
- reduces the strain on structures from the weight of water in a non-basement fire.

Incidentally, the perpetrator was later identified as a 20-year-old Canadian civilian who was then apprehended and returned to Canada.

Crash—Hercules LAPES Incident, Edmonton[13]
At 1340 hours on November 16, 1982, a Hercules C-130 transport aircraft on a training exercise crashed and burned two kilometres east of CFB Namao (Edmonton), killing all seven crewmen aboard. The plane from No. 435 Squadron was taking part in a delicate but routine manoeuvre known as Low-Level Parachute Extraction System (LAPES).

The LAPES method, as described here, has been used by both the Canadian and US Forces as a way of delivering heavy loads to areas without airstrips. In preparation for a drop, the cargo is placed on rollers in the aircraft, chocked and tied down. Two parachutes are attached to the cargo. Immediately prior to extraction the tie-downs and the chocks are removed. When drop site is reached, a drogue parachute is deployed which, in turn, draws out a large parachute that creates sufficient drag pull the cargo out of the aircraft.

On this day, just at the moment of the drop, something went wrong. The 15-ton / 13.6-tonne load did not clear the aircraft, though the "chutes" deployed behind it. The Hercules quickly assumed a nose-high attitude, stalled and plummeted into a frozen field from 600 ft / 185 m and exploded into flames. Just as the big plane hit, the load came free. It was later determined that the transfer mechanism to which the drogue chute was attached had jammed, preventing the drogue from

---

[13] *Sixty Years—RCAF and CF Air Command, 1924–1984* (1984).

pulling the extraction chutes out. This also prevented the crew from jettisoning the drogue chute. The locks were found released, and it appeared that the aluminium pallets had moved aft with an increase in pitch and power. The jammed transfer mechanism located on the ramp floor caught on the bottom of the aft pallet and prevented the load from sliding out the rear door. With the load on the ramp, the centre of gravity was thrown off and control was lost.

Ambulances, fire trucks and military vehicles raced to the scene, but there was little to do but douse the smouldering flames and recover the bodies. It was a sad day.

Fire—Aircraft Hangar, Baden, West Germany[14]
At approximately 2200 hours on March 2, 1984, a routine electronics check was being carried out on a CF 104 aircraft to confirm that the electrical circuit to the tip tanks was complete and would jettison the tanks if necessary. An important part of this check was to insert a circuit interceptor into each wing to gain the required reading, and to prevent the signal from reaching the mechanism that jettisons the tank. There were about 12 other CF104s in the hangar at the time. This building consisted of two hangar areas separated by an administrative and workshop area.

The circuit interceptor on the starboard wing was not inserted properly, therefore, when the technician activated the appropriate switch in the cockpit, the starboard tip tank jettisoned across the hangar, hitting a hydraulic test stand, broke open, ignited and sprayed its contents under an adjacent aircraft. The aircraft that precipitated the incident and an aircraft adjacent to it were almost instantly enveloped by raging fire.

The fire department was alerted and responded quickly. As the crew approached the scene, a plume of heavy black smoke was observed coming out of the only hangar door that was left partially open. The responding crew were under the command of Platoon Chief Sergeant Scotty Brown and Deputy Platoon Chief Master Corporal Bud Brown. For the initial attack they were able to nose the RIV into the partially open door and hit the fire with the turret. Broadening the attack, other firefighters advanced the handlines into the hangar and attacked the seat of the fire, while somehow avoiding exploding cannon shells. Despite intense heat, the presence of volatile fuel and exploding ordnance, the fire was quickly extinguished. All aircraft, other than the two immediately engulfed in the fire, were saved.

Of the two aircraft that were lost, the front part of one was burnt to the extent that the ejection seat had fallen to the floor, and the fuselage fuel tank located behind the cockpit was burned through to the rubber bladder. The bladder was blistered to the point that fuel was beginning to weep through the blisters. Another few seconds, and the results could have been even more catastrophic.

A Commander's Commendation in the form of a plaque was presented to the fire department by the base commander. This fire also helped to speed up a project to install a high expansion foam system that had been shelved since 1981.

Crash—Hercules Mid-Air Collision, Edmonton, Alberta[15][16]
On Friday March 29, 1985, at about 1915 hours, two C-130 Hercules collided in mid-air over the base at Lancaster Park, Edmonton, and fell to earth in a giant

---

14 Memoirs of Major Bob Giguere (1963–91) and Captain Bill Bonner (1963–95) (unpublished).
15 Material provided by the *Edmonton Journal*.
16 Memoirs of Major Barry Colledge (1970–99) (unpublished).

fire ball. Only moments before, the planes had flown over the base in a formation of three, as part of a mini-air show during a special mess dinner commemorating the 61st anniversary of the RCAF.

On board were ten military personnel—nine Canadians and one American. None survived. The two aircraft came to rest at the north end of the base. One aircraft impacted with the Long Warehouse while the second aircraft landed in the base bulk-fuel storage compound.

On duty that fateful evening were seven firefighters, the minimum staffing for CFB Edmonton, led by Warrant Officer Ross Dechaine. At the time of the crash, two firefighters were on the other side of the airfield retrieving a crash truck from Base Maintenance. Many off-duty firefighters were in the base messes that were full of personnel participating in the anniversary celebrations.

When the two aircraft collided, a "one-bell" crash alarm by the air-traffic control tower notified all that a crash had occurred. As the four firefighters in the fire hall scrambled for the crash trucks, the impact of the aircraft initiated several fire alarms and caused momentary power outages on the base.

The initial response included a Walter RIV and an Oshkosh MFV from the fire hall and a second MFV from the opposite side of the airfield. Immediately after the response was underway, the alarm room attendant started the recall of off-duty firefighters.
After colliding in mid-air, debris from the two planes scattered over a wide area as they fell to the ground. With two large aircraft down, buildings on fire, and a fire in the bulk-fuel storage compound, the scope of the incident seemed endless. One aircraft hit the ground and slid into a long wooden warehouse full of supplies. Miraculously, 150 personnel had just left the building a short time before the accident. The second plane crashed in flames into an oil storage compound, barely missing four large fuel tanks only 300 m / 328 yds away. One of the plane's tires bounced off one of the fuel tanks, leaving a huge mark on the tank. A section of one of the planes dropped into an equipment garage, filling the building with smoke. The dispatcher on duty suffered smoke inhalation and was treated in the hospital and released.

En route to the site of the crash, the platoon chief observed that one of the aircraft and a warehouse was jammed together and on fire. He sent two firefighters to the rear of the building while he proceeded to the front. The two firefighters arriving behind the warehouse discovered the second aircraft and fire in the fuel compound, which they proceeded to extinguish. This left the platoon chief with the resources of the RIV to fight the fire in the warehouse and surrounding area caused by the impact of the aircraft. Needless to say, those resources proved very inadequate for the task, since in total, one large building, one large aircraft, and six vehicles were on fire at the front of the building. Firefighters arriving at the fire hall in response to the recall quickly provided a crew for the pumper and proceeded to the scene. Thirty firefighters were on the scene within 30 minutes of the crash. Some of these were delayed due to the bottleneck created getting to and through the main gate.

The base fire department was the first-in fire department and, among other things, strove to protect the huge bulk-fuel storage tanks from fire. Firefighters from the city of Edmonton as well as the municipal district of Sturgeon responded and assisted in fighting the fire. The Alberta paramedics were on strike at the time; however, ten striking members drove out in their own cars and offered to help wherever they could. They assisted with body identification and remained at the site until their services were no longer needed.

Although the fire was under control within two hours, this team of professionals spent the next eight hours overhauling the site. By three o'clock in the morning, a skeleton crew was left to guard some stubborn hot spots and the rest of the team returned to the fire hall.

This brief description conjures a challenge for the firefighters and other emergency personnel, but provides only a minor part of the evening's activities. The tasks were a challenge, but could have been much worse since, although the large bulk-fuel storage tanks were struck with debris, only one developed a small leak, which fortunately never caught fire. One of the two firewalls in the Long Warehouse survived the impact and, with the support of some judicious fire streams, prevented the fire from reaching the radioactive materials stored on the other side. Flying debris damaged the base transport building, but none of the numerous personnel inside was injured. Some debris landed next to, but missed, the base propane facility, the 25 yd / 23 m range with small arms ammunition, the flammable-liquid drum storage, and the aircraft refuelling vehicle garage with its resident tenders.

The firefighting operations put a great demand on the base water supply made worse by leaking pipes and damaged sprinkler systems. At one point the utilities officer informed the fire chief that the reservoir was running dry and had the city boost its pressure to the maximum to help keep water in the system. The fact that the Long Warehouse was an elevated building made small hose streams ineffective, and the base electrician's cherry picker was used to provide an aerial hose stream to attack the base of the fire. Although one aircraft cockpit in free fall scattered over a large field and was cordoned off early in the operation, the other proved very difficult to locate. In fact, it wasn't until the following morning that it was found under the belly of the aircraft with very little left but ash. The other major factor was that the warehouse was being used to store materials arriving for RV 85 military exercise at Wainwright that was set to commence in April. Organizers were hard pressed to gather the required replacements.

Those who were involved in dealing with this tragic crisis will no doubt have the memory etched deeply in their minds. On the less troublesome side, they can look back and recognize those moments when the urge to do something and to do it now with extraordinary zeal was overwhelming which, in turn, resulted in happenings that, viewed in retrospect, are very revealing of the stresses that were at play. Some examples follow.

- The young driver standing on the balcony of base transport saw the aircraft heading her way, and it was rumoured she had the ambulance out of the building before the impact.
- The pumper driver parked in the path of the fire, when told to move his vehicle, did so without disconnecting from the hydrant. Fortunately, his progress was stopped quickly, but not before you could play a tune on the hose like a fiddle string.
- Late in the operation, one warrant officer led a team of firefighters into the area of the warehouse struck by the aircraft. When their hose stream contacted the heated magnesium, it erupted explosively into violent flame, causing the team to exit at a great of speed.
- A rather diminutive military policewoman discovered a photographer who had climbed the fence. She grabbed the camera he had around his neck, opened the camera, and then dragged the offender out of the fire scene.
- The base maintenance officer, dressed in his scarlet formal mess kit, was operating an old wrecker vehicle recovering vehicles from in front of the long warehouse before the fire reached them.

Combined with the grease and dirt his dress made for significant discussion.
- The base defence force that was deployed to protect the perimeter of the crash site provided some interesting moments. One individual attempted to stop a crash truck, but decided that living was more important and managed to get out of the way in time. Another tried to refuse the fire chief entry to the fire in the fuel compound because the flashlight he was carrying was not vapour-proof.
- The final note to the evening was provided by the Air Command FM when he stated that maybe now

The remains of three Hercules aircraft.

firefighters participating in Op Eval exercise scenarios would recognize that the seemingly unbelievable can, indeed, happen.

The incident took the life of ten serving members and caused over $100 million in property damage. It impaired the base's operational capabilities for several months, improved some careers and destroyed others.

After the fire was over and the recovery completed, the fire chief was asked to recognize individuals for award nominations. Although other base personnel did receive various individual awards for their actions that evening, none was given to a firefighter because the fire chief refused to nominate any individual above the team. What was received was an Air Commander's Commendation to the CFB Edmonton Fire Department on January 31, 1986.

Fire—Barge FOSS 290, Comox, British Columbia
On April 22, 1988, the *Taurus*, a Northland Service tugboat, out of the Port of Seattle, with a 61 m/200 ft barge called the FOSS 290 in tow, sailed slowly up the Straight of Georgia towards her destination in Alaska. The barge was heavily laden with general merchandise, vehicles and dangerous goods. When the tandem reached a point approximately 18 km / 11 miles east of Comox, fire was detected on the FOSS 290. Appropriate reporting procedures were followed, and at 1500 hours, the Rescue Co-ordination Centre (RCC) tasked the CFB Comox Fire Department, through Base Operations, to assess the situation and to take appropriate action.

A Labrador helicopter from 422 Squadron flew Deputy Fire Chief Master Warrant Officer Fred Johnsen, Petty Officer 2 Cliff Fueller and Master Corporal Mike Cashman to reconnoitre the burning vessel. On return to the base, Master Warrant Officer Johnsen ordered that firefighting equipment, materials and additional manpower be assembled at the Base Marine Section. Fire pumps, hose, nozzles and AFFF foam concentrate were gathered in the initial process. The motor vessel *Albatross* and crew under the direction of Petty Officer 2 Greg Saunders was used to transport the assemblage to the out-of-control blaze. Additional pumps and hose from BC Forestry were air-dropped to the scene by Buffalo aircraft.

RCC Victoria designated Master Warrant Officer Johnsen as the naval on-scene commander, an unusual appointment for an airman.

Master Warrant Officer Johnsen directed that a defensive strategic plan be implemented to prevent fire extension to fuels not already involved, to minimize the risk to firefighters and to allow for a changeover to an offensive strategic plan, should circumstance permit. To implement the plan, firefighters Petty Officer 2 Fuller, Master Corporal Cashman and Private Manczuk were ordered to board the FOSS 290 to set up pumps and to employ the defensive tactic of directing hose streams in a manner that would reduce the risk of fire spread.

The cargo manifest, provided by the captain of the *Taurus*, was reviewed and company crane operators were flown to the blazing barge from Seattle. The FOSS 290 was equipped with a crane that could be used to unload or reposition cargo containers. Concurrent activity involved the application of water and foam from the vessels Mallard and PT Race Number One. RCC Victoria deployed HMCS *Saskatchewan* from CFB Esquimalt to provide additional manpower and firefighting resources. Other vessels included the *Teal*, a Canadian Coast Guard ship, and the fire boat, *Firebrand*. On their arrival they applied water and foam in a desperate attempt to control the fire. Firefighting operations were hampered by unreliable pumps, weak hose streams, and inaccessible hot spots.

By 2000 hours, seven hours after the initial call for assistance, the fire continued to burn out of control. At 2200 hours, the crane operator arrived and he was ordered to remove a large 60000 lb / 27200 kg tank of propane to an awaiting barge. As the winds and the sea state increased, the *Taurus* reversed its course and headed southward, enabling firefighters to utilize the wind to their advantage.

At midnight, the fire was declared under control, but not totally extinguished. Overhaul and complete extinguishment could only be achieved if the firefighters could gain access to the hot spots within the containers. Master Warrant Officer Johnsen suggested overhaul operations commence at the McMillan-Blodel dock in Powell River on the east side of the Georgia Strait, more or less directly across from Comox.

Deputy Chief D. Davidson of the Powell River Fire Department was contacted and apprised of the situation.

At 0700 hours, on April 23, 1988, the fire was declared totally extinguished, some 15.5 hours after call out. Property damage estimates included four containers totally destroyed at a loss of approximately $500,000. Investigation revealed that the cause of the fire was a diesel-powered generator that provided electrical power to a refrigerator unit. General merchandise and vehicles were the main source of fuel for the fire.

The leadership and bravery of the CFB Comox firefighters and Marine Section did not go unrecognized. A Certificate of Achievement was presented to the CFB Comox Fire Department by Rear Admiral R. E. George, Commander, and Maritime Forces Pacific. The award was directed specifically to Master Warrant Officer H. F. Johnsen, Petty Officer 2 C.W. Fuller, Master Corporal M.W. Cashman and Private R. S. Manczuk.

The Marine Section was awarded a similar Certificate of Achievement.

Both Transport Canada investigators and company officials highly praised the action taken by the CFB Comox firefighters. Northland Services Incorporated, Marine Transportation, wrote, "The quick action of the military forces prevented a major disaster at sea." The *Vancouver Sun* described the event as, "Firefighters battling barge blaze." The *Powell River News* headlined, "Warm Words for the Firefighter."

Fire—City of Lahr, West Germany[17]

On September 28, 1993, at about 2220 hours, the Lahr Base Fire Department received a call on the emergency fire phone requesting assistance at a fire at Honig Un Baum tire centre in Lahr Township, adjacent to the Caserne, which housed CFE Headquarters, several children's schools, and a variety of commercial establishments, including the Lahr Exchange dry goods store. Authority was granted to proceed, and the platoon chief responded with a pumper and two MFVs. On arrival, the building was fully engulfed, and an exposure hazard was evident. The firefighting operation was being directed by the district fire chief of Baden Wurtenberg.

After consultation, the military fire chief directed firefighting operations of the Canadian contingent. Initial knockdown was achieved using the turrets of the two MFVs. This was supplemented by extending 38 mm/1¹/2 in hoselines into the building from each MFV. As well, a 65 mm/2¹/2 in supply line was laid from the Baden Wurtenberg Fire Department pumper to the DND pumper that, in turn, relayed the water through a nursing line to each MFV. Control was achieved at 2400 hours and extinguishment by 0300 hours. Once overhaul commenced, DND fire operations ceased.

---

[17] Material from CFB Lahr Fire Report and the *Lahrer Zeitung*.

Local newspapers reporting the next day stated that over 160 firefighters were involved, and that arson was suspected. The German fire brigade commander was particularly pleased about the help of the Airfield Fire Brigade. He stated, "The Canadian fire experts smothered the smoke with several thousand litres of foam. It could never have been done so quickly with our equipment."

Fire—Dartmouth, Nova Scotia
The night of Sunday, January 19, 1997, was quiet in the sprawling Dartmouth, Nova Scotia, apartment complex at 7 Jamieson Street, a stone's throw from Halifax Harbour. Outside it was bitterly cold and snowy. In Apartment 324 at about 1 a.m., Dave Crocker went to bed.

Across the hall in Apartment 321, Wayne Wells had long been in bed. He had to leave for work at 6 a.m. and before that he generally took Bear, the family's big Australian shepherd, for a walk. His wife Debra went to bed shortly after midnight, admonishing her 17-year-old daughter Jamie, who was babysitting her two-and-one-half-year-old cousin Donald Joseph, not to stay up too late.

Dave arose around 7:30 the next morning, and at about 9:20, he smelled something peculiar. Glancing up at his front door, he saw with alarm that smoke was snaking around the edges of the frame. At that moment, fire bells went off throughout the building. Dave opened his door into the hallway and saw smoke coming from Apartment 321. He pounded on the door to this apartment and heard someone say, "There's a fire in here!" There was no response to his call to open the door.

Like all doors in the building, this was a steel-clad fire-separation door. Without an axe, there was no way to open it. Dave returned to his apartment and dialled 9-1-1 to report the fire. The operator asked him stay on the line, but he replied, "I can't, there are people who need help." With that, he hurried to the superintendent's flat to get the key to open the apartment. When he arrived, he was informed that the superintendent had gone to alert residents and had taken the keys with him. He then retraced his steps.

In her bedroom, Jamie awakened slowly, wondering if she heard her mother call or if had she been dreaming; DJ lay sleeping beside her. Curious, she walked to her bedroom door and opened it. Black smoke billowed into the room, blinding her. In seconds, the room was so dark that she could barely make out the bed. Numbed with terror and gagging, the asthmatic teenager groped for her sleeping cousin. Gathering DJ in her arms, she began feeling her way desperately towards what she hoped was the front door. She had progressed no more than 3 m / 10 ft when the smoke overwhelmed her. Her lungs burning, Jamie slumped against the wall while holding DJ and slid to the floor.

On the verge of complete collapse, Jamie felt something warm and wet seize her ankle. Reaching down, she realized it was Bear. The dog could see under the dense smoke and was gently but firmly tugging at her with his big mouth. She forced herself to stand up and then let herself be guided by the dog to the front door where she found and released the dead-bolt lock. Suddenly the door swung open. Amid a rush of black smoke, soot-stained Jamie, led by Bear, emerged from the apartment. As Jamie, blinded by the smoke, made her way down the hallway towards the stairs, she was met by Dave who carried her and DJ towards the stairs. As they approached the stairs, neighbours reached out to take Jamie and DJ down to safety. The girl said, "My mother's still in there!"

Dave returned to the apartment and this time had to crawl on his hands and knees. At the rate the smoke was lowering, there would be no breathable air in a matter of minutes. As he struggled forward, eyes stinging, Dave tried to formulate a strategy for finding the girl's mother. Would she be in one of the bedrooms? If so, where were they? He knew 321 was a two-bedroom apartment. He wondered if the layout was similar to his own one-bedroom unit. There appeared to be a dividing wall to his left—the kitchen? If so, perhaps the bedrooms were down what looked like a corridor to his right. "Is there anybody in here?" he called out. He was about to start off to his right when he heard a moan on his left. Straining to see, he perceived a slight movement and crawled towards it. An object came into focus—a woman's foot. Dave hustled towards her. Grabbing her shoulders, he pulled hard, but she didn't move. It was too dark to see beyond the woman's head, so he moved forward. Then he saw that the woman was gripping the edge of the dividing wall. Dave yelled, "Let go! You've got to let go!" and managed to free her hands. Then, taking the semiconscious woman firmly by the arms, he pulled her towards the door.

The deadweight of the victim proved far heavier than Dave had anticipated. He had to get up on his knees and heave to move her. In doing so, he momentarily thrust his head up into the lowering ceiling of smoke and felt the intense heat and realized that the fire could "flash-over" any second. With Debra in tow, he reached the door and the relative safety of the hallway just as an explosion occurred. Neighbours took her outside, and he returned to search for the dog. By now the smoke was close to the floor. He had to slither along on his belly in order to breathe. "Here, boy! Here, boy!" he called but there was no response. The apartment was now completely ablaze. If any living thing was still in there, it was too late to do anything about it.

There was now a lot of smoke in the hallway, so Dave returned to his apartment and closed the door. Smoke now filled his place, too. If the blaze jumped the hallway, he would have to act quickly to protect his property. Some smoke was coming into the apartment, so he opened the balcony doors and turned on the fans over the kitchen stove for ventilation and soaked towels in water and put them against the bottom of his door. As he did so, he heard heavy boots tramping through the hallway. Someone pounded on his door and shouted, "Fire department! Anybody in there?"

"I'm okay!" Dave yelled back. "I've got towels around the door. Don't break it in!" Then he went to the balcony to breathe in fresh air. Looking up, he saw that the roof was on fire!

Dave considered his situation: The apartment complex was built on the side of a hill, so his balcony was only two storeys up. He could, he reckoned, climb down to the balcony below his and jump. At most he'd break a leg. As he was deliberating what he should do, firefighters came around the back of the building and placed a ladder against the balcony railing. As a firefighter started up, Dave yelled down, "It's okay. I'm a fireman. I can climb down myself!"

Dave Crocker went to visit Debra Wells in hospital a week later. She did not remember a thing about the incident, other than calling to her daughter that there was a fire. "What do I say to the guy who saved my life?" she asked him. As for the dog, it was found that he had slipped away unseen after he guided Jamie and DJ to the front door. He now lives in the country with a friend of the Wells family.

On March 5, 1998, Dave Crocker was awarded the Carnegie Hero Fund Commission's Medal for Extraordinary Heroism, and on September 17, the Governor General's Award for Bravery. He also became the first DND firefighter to win Canada's Firefighter of the Year Award.

Fire—The Gijon Fishing Trawler, Esquimalt, British Columbia[18]

On October 27, 1997, a huge fire engulfed the *Gijon*, a stern trawler fishing vessel of Russian registry, while it was inside the Esquimalt graving dock. In all, seven fire departments responded, including the Township of Esquimalt; the Municipalities of Victoria, Sannich, Colwood, Langford, and View Royal; and HMC Esquimalt Dockyard.

The total number of apparatus responding from the municipal departments is not known; however, it is known that DND dispatched a pumper, a ladder truck, a HAZMAT vehicle and a fire boat. The fire raged for over 65 hours, and it was estimated that the total manpower involved exceeded 100 firefighters

The firefighters first on the scene attempted an onboard attack, but they were forced back by the extreme heat and the danger from exploding gas cylinders. The gangway was lost soon after, leaving no way to board or to exit the ship. The defensive attack included large volumes of water to cool down the ship. In addition, foam was applied for a considerable time; however, local authorities were afraid of polluting the harbour, which left water as the only other option. Only after the fire was out did the firefighters learn that 2500 lb / 1135 kg of the highly hazardous compound sodium hydroxide—a white, deliquescent, water-soluble caustic solid that upon the introduction of moisture liquefies and generates heat—was stored onboard. Fortunately it was stored in a water-tight compartment and stayed dry.

Excluding wild fires where loss is difficult to quantify, the *Gijon* fire, which took two-and-a-half days to extinguish, remains the costliest fire in British Columbia's history. DND was subsequently involved in a claim against the trawler's owners for the sum of US $100 million.

In June 2003, all disputes were resolved and a settlement was reached between the two parties. DND was cleared of all allegations, claims, and demands against them. The settlement was reached out of court; consequently, no information was made public.

Summary

These incidents typify the professionalism and spirit of the members of our Fire Service. They have repeatedly demonstrated the knowledge, skill, and guts to deal with the demanding unpredictability of emergencies. Who amongst us could ever have predicted the situations faced by Corporal Neufeld, Firefighter 1 Crocker, and many others? Surely no one could. That these situations and others like them have been confronted with such distinction is a valued testimonial.

---

8 Personal Memoirs of Firefighter 7 Paul Beaulieu (1966–04) and Captain Ken Hoffer (1969–). Unpublished.

IGNI OBSTARE

BIBLIOGRAPHY

Conrad, Peter C. — *Training for Victory, the British Commonwealth Air Training Plan in the West.*

Douglas, W.A.B. — *The Creation of a National Air Force.* University of Toronto Press. Government Catalogue No. D2-63/2-1985E. 1986.

Hatch, F. J. — *Aerodrome of Democracy: Canada and the British Commonwealth Air Training Plan, 1939–1945.* Department of Defence Directorate of History. Canadian Government Publishing Centre. Catalogue No. D63-1-3F 1983.

Shores, Christopher — *History of the Royal Canadian Air Force.* Bison Books Corporation W.H. Smith Publishing Canada. 1984.

Williams, James N. — *The Plan: Memories of the British Commonwealth Air Training Plan.* Williams John Deyell Publishing Company. 1984.

Wise, S.F. — *Canadian Airmen and the First World War.* University of Toronto Press. 1980.

— *Army Fire Manual.* Issued with Army Orders, July 1, 1917.

— *Fire Manual for Hutment Camps.* Issued with Army orders, July 1, 1916. DND Directorate of History File No: 86/172.

— *Fire Services Manual for the Canadian Army 1948.* DND Directorate of History File No. 91/158.

— *History of the Canadian Corps of Firefighters.* National Archives of Canada Record Group 38, Vol. 141. (Obtained from Veterans Affairs).

— "Memorandum: Department of National Defence 1700-139 Vol. 1 to Treasury Board 3rd November 1948." (Request to increase the manpower levels of the Navy to accommodate firefighter positions.)

— "Memorandum: Defence Services Hon. Brooke Claxton Establishment of Firefighters for the Defence Department, 8 November 1948." Request to have all firefighter positions within DND be entirely comprised of servicemen.

— "Minute: No 362827B Treasury Board Ottawa 19 January 1949." The minimum number of firefighters essential for the efficient operation of the fire services.

— "Regulations for Army Fire Services 1934." DND Directorate of History, File No. 86/172.

Appendix A

Key Appointments 1938–2004

INTRODUCTION
In the process of compiling this list of key appointments, it became increasingly evident that some inaccuracies and oversights would have to be accepted. To hold otherwise would require that the project be abandoned. It was on the basis of this precept that it was assembled.

COMPOSITE TRAINING SCHOOL—OFFICERS IN CHARGE
RCAF Station Mountain View
Flying Officer W.A. McCallum	1941–42
Pilot Officer T.H. Mathews	1942–42

RCAF Station Trenton
Flight Sergeant Art McFayden	1942–43
Flying Officer B.C. Beazer	1943–44

FIRE SCHOOL SUPERVISORS—OFFICERS COMMANDING
RCAF Station Trenton
Flight Sergeant Art MacFayden	1946–47
Flight Sergeant Jock Smith	1947–48
Flight Sergeant Art MacFayden	1948–50
Warrant Officer 1 Dave Lefebvre	1950–51

RCAF Station Aylmer
Warrant Officer 1 Dave Lefebvre	1951–54
Warrant Officer 1 Bob Edwards	1954–56

RCAF Station / CFB Camp Borden
Flying Officer Art McFayden	1956–59
Flight Lieutenant Bill Maggs	1959–62
Flight Lieutenant Phil Brown	1962–67
Flight Lieutenant Gord Ferris	1967–71
Flight Lieutenant Art Haggart	1971–75
Flight Lieutenant Gord Ferris	1975–77
Captain Bob Maxwell	1977–80

Major Jim Brook	1980–84
Major Charles McNeil	1984–85

COMMANDANT—CANADIAN FORCES FIRE ACADEMY (Borden, Ontario)

Major Charles McNeil	1985–90
Major Bob Giguere	1990–91
Major John Gordon	1991–95
A/Major Denis Girouard	1995–97
Major Barry Colledge	1997–99
Major Tony Lovett	1999–2000
Major Jim Murray	2000–03
Major Sean Lewis	2003+

NAVY FIRE MARSHALS[1] (Ottawa, Ontario & Halifax, Nova Scotia)

Commander C.A. Thomson	1940–45
Lieutenant Commander E.F. Bevis	1945–47
Lieutenant Commander W. Simpkin	1948–60
Lieutenant Commander Neil Duval	1960–67
Mr. Gordon Lay	1967–79*
Mr. Jim McInnis	1979–88
Major John Gordon	1988–91
Major Andre Beaudin	1991–93
Lieutenant (N) Chris Derossenoll	1993–98
Captain Neil Drachenberg	1998–99
Captain Ken Hoffer	1999+

* There was a gap here during the regional concept.

NAVY—ATLANTIC COMMAND (Halifax, Nova Scotia)

Lieutenant C. Spinney	1941–44
Lieutenant Commander Edgar Bevis	1944–44
Lieutenant W.J. Carson	1944–45
Lieutenant Commander J. Harber	1946–53
Lieutenant H. Curran	1953–54
Lieutenant Gordon Lay	1954–65

NAVY—PACIFIC COMMAND (Esquimalt, British Columbia)

Lieutenant Commander J.D. Crowther	1941–56
Lieutenant Commander Neil Duval	1956–59

---

[1] The title changed to Maritime Command in 1968 and to Chief of Maritime Forces in 1997. The regional concept was in effect at least through 1968–70.

Lieutenant Commander N. Stewardson	1959–63
Lieutenant Commander Gordon Ball	1963–1966
Lieutenant Commander Alex Hope	1966–**

** The length of the appointment was not determined.

ARMY FIRE MARSHALS[2] (Ottawa, Ontario & St. Hubert, Quebec)

Major P.C. Ahern	1939–42
Lieutenant Colonel O.L. Lister	1942–43
Major A. DesRoche	1943–45
Lieutenant Colonel E. Desjardins	1945–60
Lieutenant Colonel Lindsey-Brown	1960–66
Squadron Leader Bill Walker	1966–67
Flight Lieutenant George Palmer	1967–68
Mr. Gerry Berube	1970–85
Mr. Stoney Bourque	1985–95
Captain Don McNeil	2001+

AIR FORCE FIRE MARSHALS[3] (Ottawa, Ontario & Winnipeg, Manitoba)

Flight Lieutenant P.S. Snarr	1938–41
Squadron Leader R.E. Ritchie	1941–45
Squadron Leader W.D. Martin	1946–47
Flying Officer Bob Armour	1948–50
Flying Officer Bert Quinn	1950–52
Wing Commander Bill McCallum	1952–58
Wing Commander Bert Quinn	1958–66
Major Lorne MacLean	1975–78
Major Hal Singleton	1978–82
Major Charlie MacNeil	1982–85
Major Bob Giguere	1985–90
Major Garry Mauch	1990–1994
Lieutenant (N) Wayne Graves	1994–96
Captain Jeff Carlisle	1996–98
Captain Gaetan Morinville	1998–2001
Captain Allan Rau	2001+

---

[2] The title changed to Mobile Command in 1968 and to Chief of Land Forces in 1997. During a transitional period, 1995–2001, the AFM responsibilities were exercised by a desk officer in the CFFM's office. The regional concept was in effect at least through 1968–70.

[3] The title changed to Air Command in 1975. Essentially, there was no AFFM, per se, between 1966 and 1975. CFMs were in place for Defence Command, Transport Command, and Training Command. The regional concept was in effect at least through 1968–70.

EASTERN AIR COMMAND FIRE MARSHALS
Flight Lieutenant W.A. Carlisle	1941–43
Flight Lieutenant L.J. Bishop	1943–44
Flight Lieutenant T.H. Mathews	1944–44
Flight Lieutenant B.C. Beazer	1944–45
Warrant Officer 2 Dave Lefebvre	1945–46
Warrant Officer 2 Paul Palylyk	1946–47

WESTERN AIR COMMAND FIRE MARSHALS
Flight Lieutenant F.C. Eldred	1941–44
Flight Lieutenant W.A. Carlisle	1944–46
Warrant Officer 2 Fred Sacho	1946–47

NO. 1 AIR TRAINING COMMAND FIRE MARSHALS
Flight Lieutenant L.J. Bishop	1940–43
Flight Lieutenant R.J. Mathieson	1943–45

NO. 2 AIR TRAINING COMMAND FIRE MARSHALS
Flight Officer G.E. Huff	1941–42
Flight Lieutenant W.A. MacCallum	1942–46

NO. 3 AIR TRAINING COMMAND FIRE MARSHALS
Flight Lieutenant R.J. Mathieson	1942–43

NO. 4 AIR TRAINING COMMAND FIRE MARSHALS
Flight Lieutenant T. Jenkins	1941–45

NORTHWEST AIR COMMAND FIRE MARSHALS
Flight Sergeant F.A. Philips	1946–46

RCAF Air Defence Command*
(St. Hubert, Quebec & North Bay, Ontario)
Flight Lieutenant Wally Sinclair	1956–57
Flight Lieutenant John Cowell	1957–60
Squadron Leader Bill Walker	1960–65
Captain Art Haggart	1965–68
Major Gerry Torraville	1968–73

RCAF TRAINING COMMAND*
(Winnipeg, Manitoba)
Flight Lieutenant Jules Debrouvers	1960–63

Mr. Ferguson	1963–70
Major Phil Brown	1970–74
Captain Don Fulmer	1974–75

RCAF TRANSPORT COMMAND* (Trenton, Ontario)
  Flight Lieutenant Wally Sinclair   1954–57

RCAF MARITIME AIR COMMAND* (Shearwater, Nova Scotia)
  Flight Lieutenant Hank Webster   1963–66
  Flight Lieutenant Bill Maggs   1966–69

RCAF WESTERN AIR COMMAND (Edmonton, Alberta)

RCAF NUMBER 5 AIR GROUP (Vancouver, British Columbia)
  Flight Lieutenant Leo Herman   1963–67

* The interruptions associated with the regional concept were not positively identified.

ATLANTIC REGION FIRE MARSHAL (Halifax, Nova Scotia)
  Mr. Gordon Lay   1968–70

QUEBEC REGION FIRE MARSHAL (St. Hubert, Quebec)
  Mr. Gerry Berube   1968–70

ONTARIO REGION FIRE MARSHAL (Ajax, Ontario)
  Mr. Buck Buchannan   1968–70

WESTERN REGION FIRE MARSHAL (Edmonton, Alberta)
  Details not available.   1968–70

PACIFIC REGION FIRE MARSHAL (Vancouver, British Columbia)
  Details not available.   1968–70

Note: The dates for the regional concept were not positively established.

NO. 1 AIR DIVISION FIRE MARSHALS (Metz, France)
  Flight Sergeant Bill Maggs   1952–54
  Flying Officer Archie Graham   1954–57
  Flight Lieutenant Wally Sinclair   1957–60

Squadron Leader Bill Walker	1960–64
Flight Lieutenant John Cowell	1964–66

TRAINING SYSTEMS FIRE MARSHALS (Trenton, Ontario)

Major Art Haggart	Unknown
Captain Bill Dawson	*–1976
Major Marty Jack	1976–79
Major Charlie McNeil	1979–82
Captain Gord Gazley	1982–85
Major John Gordon	1985–88
Captain Jim Wright	1988–89
Lieutenant (N) Chris Derossenoll	1989–93
Captain Peter Keough	1993–95

AUXILIARY EQUIPMENT PROCUREMENT OFFICERS

Warrant Officer 2 Fred Sacco	1952–54
Flying Officer John Cowell	1954–57
Flying Officer Archie Graham	1957–60
Flying Officer Hank Webster	1960–63
Flying Officer Doug Stevenson	1963–64
Warrant Officer 1 Jack MacDougall	1964–*
Warrant Officer 1 Bobby Edwards	*–1974
Master Warrant Officer Jim Landry	1974–77
Chief Warrant Officer Jack Kroeplin	1977–81
Chief Warrant Officer Livingston	1981–85
Chief Warrant Officer "Red" Ash	1985–89
Chief Warrant Officer Jim Munro	1989–93
Chief Warrant Officer Ed Neufeld	1993–96
Chief Warrant Officer Jim Munro	1996–98
Chief Warrant Officer Ken Hoffer	1998–2000
Chief Warrant Officer Steve Shand	2000–04
Chief Warrant Officer John Barker	2004+

CAREER MANAGERS (Ottawa, Ontario)

Warrant Officer 2 Gord Shaw	*–1967
Warrant Officer 2 John Kimick	1967–70
Warrant Officer Joe Gilliland	1970–74
Chief Petty Officer 2 Buster Noyes	1974–77

*Dates were not determined.

Chief Petty Officer 2 Bob Connors	1977–80
Master Warrant Officer Joe Walker	1980–84
Chief Petty Officer 2 John Daley	1984–86
Master Warrant Officer Brian Buchanan	1986–89
Master Warrant Officer Don Adlam	1989–91
Chief Warrant Officer Jim Munro	1991–95
Chief Warrant Officer Bob Morrison	1995–2003
Chief Warrant Officer Roger Gosteau	2003+

CANADIAN FORCES EUROPE FIRE MARSHALS (Lahr, West Germany)

Flight Lieutenant Keith Potter	1966–69
Captain Lorne MacLean	1969–73
Captain George Cowan	1973–77
Captain Don Carmichael	1977–80
Captain Bob Giguere	1980–84
Captain Dave Geddes	1984–87
Major Garry Mauch	1987–90
Major Charlie MacNeil	1990–92
Captain Barry Colledge	1992–93

CANADIAN FORCES FIRE MARSHALS (Ottawa, Ontario)

Lieutenant Colonel Lindsey-Brown	1966–67
Commander Neil Duval	1967–68
Commander Gordon Ball	1968–69
Lieutenant Colonel Ken Chisholm	1969–72
Commander Alex Hope	1972–73
Lieutenant Colonel Gerry Torraville	1973–76
Lieutenant Colonel Bill MacDonald	1976–78
Lieutenant Colonel Lorne MacLean	1978–86
Lieutenant Colonel Hal Singleton	1986–91
Lieutenant Colonel Bob Maxwell	1991–93
Acting Lieutenant Colonel Gaetan Perron	1993–95
Lieutenant Colonel Stew Moore	1995–97
Lieutenant Colonel Sean Tracey	1997–2000
Lieutenant Colonel Tony Lovett	2000–02
Lieutenant Colonel Marc Desjardins	2002–2005
Major Gaetan Morrinville	2005

Appendix B

Descriptions & Definitions

AID
A nuclear weapon accident, incident, or deficiency (AID) is a condition that degrades nuclear surety/safety. This includes damage, loss, destruction, malfunction, or exposure to adverse weather conditions of any part of the weapon system.

Air-Aspirating Playpipe Nozzles
This type of nozzle has an extended pipe-like barrel, hence the name. After the foam concentrate / water mixture passes through the air-aspirating feature of the nozzle, it expands to the point where it lacks the density to maintain its sense of direction without the guidance of the extended barrel.

Bent Spear
The term used to identify:
- a nuclear-weapon system incident;
- a significant incident or unexpected event involving a nuclear weapon, warhead, or nuclear component resulting in damage to the extent that major rework, examination, re-certification, or complete replacement by the design agency is required; or
- an incident requiring immediate action in the interest of safety or nuclear weapons security which may result in adverse national or international public reaction.

Bakelite
Bakelite is a brand name for any of a series of thermosetting plastics prepared by heating phenol or cresol with formaldehyde and ammonia under pressure. It is used for radio cabinets, telephone receivers, electric insulators, and moulded plastic ware.

Bar
A unit of pressure equal to approximately atmospheric pressure (100 kPa / 14.7 psi).

Broken Arrow
The term used to identify:
- a nuclear-weapon system accident; or
- an unexpected event involving a nuclear weapon, warhead, or nuclear component resulting in destruction of a nuclear weapon, radioactive contamination, or a hazard to the public either actual or implied, such as nuclear or non-nuclear detonation of a weapon.

Burnback
The heat-resistance property of expanded foam to withstand exposure to high heat fluxes without loss of stability that would permit a fire to expand over areas previously secured. This expansion of fire is commonly referred to by the term burnback.

Cold War
Intense economic, political, military, and ideological rivalry between nations, short of military conflict; sustained hostile political policies and an atmosphere of strain between opposed countries. The implications of the term, insofar as it is used in this document, relates to the rivalry that developed, after World War II, between the Soviet Union and its satellites and the democratic countries of the Western World, under the leadership of the United States.

Demand-Type SCBA
This type of SCBA has a valve between the air tank and the face piece designed so as to deliver air to the wearer in response to negative pressure in the face piece caused by the wearer inhaling. The critical deficiency in this operating principle was that the negative pressure in the face piece could result in contamination gaining entry.

Dull Sword
The term used to identify:
- a nuclear-weapon system deficiency;
- a situation, event, or condition not reportable as a BENT SPEAR or BROKEN ARROW, which could or does degrade nuclear surety/safety;
- damage to, or malfunction of, a nuclear weapon, component, or associated equipment;
- exposure of a weapon, warhead, or critical component to severe or unusual environments;
- unplanned, unexpected, or inadvertent release, launch, loss, or destruction of a nuclear- training device;
- damage to nuclear-certified, ground-support equipment, including common commercial vehicles and munitions handling equipment; or
- damage, failure, or malfunction of a nuclear-capable delivery vehicle, suspension, release, launch, separation, arming, monitoring, or control systems;

Foam
Foam, as the term is used to describe a fire-suppressing agent, is an aggregation of small bubbles used to form an air-excluding, vapour-suppressing blanket over the surface of a flammable liquid or other fuel.

Foam, Alcohol-Resistant
Alcohol-resistant foam is used for fighting fires involving water-soluble materials or fuels that are destructive to other types of foams. Some alcohol-resistant foam may be capable of forming a vapour-suppressing aqueous film on the surface of hydrocarbon fuels.

Foam, Aqueous Film Forming (AFFF)
A concentrated aqueous solution of one or more hydrocarbon or fluorochemical surfactants that forms foam capable of producing a vapour-suppressing aqueous film on the surface of hydrocarbon fuels.

Foam Concentrate
A concentrated liquid foaming agent that is mixed with water and air in designated proportions to form bubbles that constitutes expanded foam.

Foam Drainage Time
Generally used as a measurement of the time in minutes that it takes for 25% of the total liquid contained in the foam sample to drain from the foam. Often referred to as the quarter life of the foam.

Foam Expansion Ratio
The ratio between the volume of foam produced and the volume of foam solution used in its production.

Foam, Protein
A protein-based foam concentrate stabilized with metal salts to make a fire-resistant foam blanket.

Foam Solution
The result when a foam concentrate and water are mixed in designated proportions, prior to aerating to form expanded foam.

Ghost Flame
Flames that rise, flicker and seemingly disappear only to rise again, perhaps then to expand rapidly across the surface of a liquid fuel.

Homelite Saw
Powered by a gasoline engine, the Homelite Saw was hand-portable and had a variety of blades with different tooth designs to allow it to be used for cutting wood, metal, concrete, or other materials. Its uses were mainly for cutting openings for ventilating a burning building or in rescue work.

Human Reliability Program
A process of evaluating individuals, beyond the normal security clearance procedures, to assess the individual's propensity for committing an irrational act that could pose a risk to a nuclear weapon.

Hydro-Blender
The term, as used in this document, refers to equipment that combined water with another compound to reduce the surface tension of the water so as to cause the ensuing mix to be more effective in suppressing fire in combustible materials, where penetration of the liquid into the material was needed to achieve extinguishment. The mix was commonly referred to by the trade name of Wet Water. The units used on fire apparatus, and in some cases as part of

fixed installations in buildings, featured a tubular steel or plastic cylinder approximately 4 in / 10 cm in diameter and 12 in / 30 cm to 18 in / 45 cm in height with an inlet and outlet for water to pass through under pressure. The compound that was added to reduce the surface tension of the water was in the form of a glycerine capsule. The operating principle was that water swirling around in the hydro-blender, en route to the discharge point, would entrap other compounds from the capsule and thereby achieve the desired result.

Kevlar
A trade name for aramid, any of a class of long-chain polyamides capable of extrusion into fibres having resistance to high temperatures and great strength.

Kilopascal
One thousand pascals.

Mystery Nozzle
With two main parts, the inner part attaches to the hose and the outer part envelops the inner part like a sleeve and can be manually rotated. This rotation changes the discharge pattern through an innumerable number of ranges from straight stream to the point where it radiates in a circular pattern that is virtually 90° off the plane of the face of the nozzle, similar to the fanned tail feathers of a strutting male peacock.

Newton
The SI unit of force, equal to the force that produces an acceleration of one metre per second per second on a mass of one kilogram.

No Lone Zone
An unmarked yet rigidly defined peripheral boundary surrounding a nuclear weapon within which the entry of a single person is prohibited.

Nucflash
The term used to identify an accidental or unauthorized launching, firing or use of a nuclear weapon or weapons system by US or non-US supported allied forces which could cause the outbreak of war.

Pascal
The SI unit of pressure or stress, equal to one newton per square metre.

Pressure-Demand SCBA
Has a valving arrangement between the air tank and the face piece designed so as to maintain positive pressure in the face piece to preclude the entry of contamination.

Sally Port
Dating back to the Middle Ages, this secret passageway was designed to let one person enter or leave the castle at a time. It was a complicated tunnel defended with murder holes and a portcullis to make sure the castle had no unwelcome visitors.

The concept, without the murder holes and such, was adapted as a means of controlling the movement of vehicles entering or exiting SAS depots. They were constructed in the form of a large rectangular pen surrounded with security fencing fitted with gates at each end for vehicles and a gate on one side to channel persons into a guard house for pass exchange and other security checks. The gates were remotely controlled from the guardhouse. Procedurally, when a vehicle approached the outer gate, it would, given normal circumstances, be opened by security personnel. The vehicle would then enter the sally port after which the outer gate would close, effectively impounding the vehicle and its occupants. With all gates now in the closed position, the second gate would remain closed until documentation was cleared, after which the inner gate would be opened to allow the vehicle to proceed. The same procedures applied when exiting a SAS.

Scaling
The assessment of each building against a predetermined scale of issue for portable fire extinguishers and obtaining and installing the required equipment.

Special Armament Storage
Facilities primarily for the storage of nuclear munitions. They were designed to satisfy the overriding principle of having security take precedence over all other considerations.

SITREP
A situation report to a command post, or other monitoring organization, of conditions at any given point during emergency operations.

Snozzle
The term was coined by Crash Rescue Equipment Service, Inc., of Dallas, Texas, to identify a device designed with the aim of improving the firefighting capabilities of fire apparatus. Essentially, it is an articulated and elevated water tower controlled by the vehicle operator using a cab-mounted joystick. It is a very versatile tool that, when fitted to an ARFFV, is capable of delivering PK, water and foam, and is equipped with a nozzle, camera and a floodlight. The nozzle can pierce an aircraft fuselage and penetrate into the interior to apply fire-suppressing agents.

Theoretical Critical Area
Adjacent to an aircraft in which fire must be controlled for the purpose of ensuring temporary fuselage integrity and providing an escape area for its occupants. The TCA is a rectangle, having as one dimension the overall length of the aircraft, and the other dimension determined by the following:

- for aircraft with an overall length of less than 20 m / 65 ft, 12 m / 40 ft plus the width of the fuselage; and
- for aircraft with an overall length of 20 m / 65 ft or more, 30 m / 100 ft plus the width of the fuselage.

The TCA serves only as a means for categorizing aircraft in terms of the magnitude of the potential fire hazard in which they may become involved.

Vehicle Code

For some vehicles, this is expressed by letter in upper case followed by a number, without any spacing between the characters, e.g. G8.

Wildfire

An uncontrolled fire, usually spreading through vegetative fuels and often threatening structures.

Appendix C

Acronyms & Abbreviations

Abbreviations—General

B	bar
Cdn	Canadian
cm	centimetre
ft	one linear foot
ft^2	one square foot
ft^3	one cubic foot
G	Imperial gallon
G/min	Imperial gallons per minute
hp	horse power
kg	kilogram
kPa	kilopascal
kph	kilometres per hour
L	litre
lb	pound
L/min	litres per minute
m	metre
m^2	square metre
m^3	cubic metre
mph	miles per hour
mm	millimetre
NR	not recorded
psi	pressure in pounds per square inch

qt	quart
s	second
SI	International System of Units (weights and measures)
USG	United States gallon
USG/min	United States gallons per minute
yd	yard, which has a lineal measurement of 3 ft / 0.9 m.

Abbreviations—Rank Titles

AB	Able Seaman
AC	Aircraftsman
Adm	Admiral
BGen	Brigadier-General
Capt	Captain
Cdr	Commander
Cmdre	Commodore
Col	Colonel
Cpl	Corporal
CPO 1	Chief Petty Officer, 1st Class
CPO 2	Chief Petty officer, 2nd Class
CWO	Chief Warrant Officer
F/L	Flight Lieutenant
F/O	Flying Officer
FR	Firefighter
F/S	Flight Sergeant
FSO	Fire Service Officer
G/C	Group Captain
Gen	General
LAC	Leading Aircraftsman
LCdr	Lieutenant-Commander
LCol	Lieutenant-Colonel
LGen	Lieutenant-General
LS	Leading Seaman
Lt	Lieutenant
2Lt	Second-Lieutenant

Maj	Major
MCpl	Master Corporal
MGen	Major-General
MWO	Master Warrant Officer
OS	Ordinary Seaman
PO 1	Petty Officer, 1st Class
PO 2	Petty Officer, 2nd Class
Pte	Private
RAdm	Rear Admiral
Sgt	Sergeant
S/L	Squadron Leader
SLt	Sub-Lieutenant
S/S	Staff Sergeant
VAdm	Vice-Admiral
W/C	Wing Commander
WO	Warrant Officer
WO1	Warrant Officer, 1st Class
WO2	Warrant Officer, 2nd Class

Acronyms

ACFM&FC	Association of Canadian Fire Marshals and Fire Commissioners
ADD	Anti-Discrimination Directorate
ADHQ	Air Division Headquarters
AFFF	Aqueous Film-Forming Foam
AFFM	Air Force Fire Marshall
AFHQ	Air Force Headquarters
AFM	Army Fire Marshall
AGM	Annual General Meeting
AID	Accident, Incident, or Deficiency
AOC	Air Officer Commanding
ARF	Alcohol-Resistant Foam
ARFF	Aircraft Rescue and Firefighting
ARFFV	Aircraft Rescue and Firefighting Vehicles
ASME	American Society of Mechanical Engineers
BCATP	British Commonwealth Air Training Plan
BEM	British Empire Medal

BFOR	Bona Fide Occupational Requirements
BMFD	Bedford Magazine Fire Department
CAF	Canadian Armed Forces
CAFC	Canadian Association of Fire Chiefs
CAS	Chief of the Air Staff
CBG	Canadian Brigade Group
CBM	Clorobromomethane
CCFM&FC	Canadian Chief of Fire Marshals and Fire Commissioners
CD	Canada Decoration
CDS	Chief of Defence Staff
CE	Construction Engineering
CEO	Construction Engineering Officer
CF	Canadian Forces
CFAO	Canadian Forces Administrative Officer
CFAT	Canadian Forces Aptitude Test
CFB	Canadian Forces Base
CFE	Canadian Forces Europe
CFFA	Canadian Forces Fire Academy
CFFM	Canadian Forces Fire Marshal
CFIS	Canadian Fire Investigation School
CFM	Command Fire Marshall
CFPO	Canadian Fire Prevention Officer
CFR	Commissioned from the Ranks
CFRP	Commissioning from Ranks Plan
CFSAOE	Canadian Forces School of Aerospace, Ordnance and Engineering
CFSIT	Canadian Forces School of Instructional Technique
CFTS	Canadian Forces Training System
CHRC	Canadian Human Rights Commission
CI	Capability Inspection
CIB	Canadian Infantry Brigade
CIBG	Canadian Infantry Brigade Group
CJS	Canadian Joint Staff
CME	Canadian Military Engineers
CO	Carbon Monoxide
CO_2	Carbon Dioxide
CPAO	Civilian Personnel Order
CRA	Compulsory Release Age
CRAD	Chief of Research and Development
CSA	Canadian Standards Association

CSTF	Canadian Standardized Test of Fitness
CT	Control Time
CTC	Carbon Tetrachloride
DC	Sodium Bicarbonate Dry Chemical
DelegAAT	Delegation of Authority and Accountability Trial
DEW	Distant Early Warning
DMC	Defence Management Committee
DND	Department of National Defence
DPERA	Directorate of Physical Education and Recreation
DRDHP	Directorate of Research and Development Human Performance
EDITH	Exit Drills in the Home
EXPRES	Exercise Prescription Plan
4F	Grade 4, fit for duties of light or sedentary nature
FF	Posted by Authority of the Director of Fire Safety
FFFPF	Film-Forming Fluoroprotein Foam
FFSPFP	Firefighter Specific Physical Fitness Program
FFTC	Firefighter Training Company
FPC	Fire Prevention Canada
FPF	Fluoroprotein Foam
FSO	Commissioned Fire Service Officer
FSOC	Fire Service Officer Corps
GD	General Duties
HAZMAT	Hazardous Materials
HMCS	His/Her Majesty's Canadian Ship
HRP	Human Reliability Program
IFSAC	International Fire Service Accreditation Congress
IFSTA	International Fire Service Training Association
JFMC	Joint Fire Marshals Committee
JNBC	Joint Nuclear Biological and Chemical
JSFC	Joint Services Fire Committee
KTS	Composite Training School

LAC	Leading Aircraftsman
LAPSE	Low-Level Parachute Extraction System
LRV	Light Rescue Vehicle
MFV	Major Foam Vehicle
MM	Military Medal
MMM	Member, Order of Military Merit
MOC	Military Occupation Code
MP	Military Police
MPFS	Minimum Physical Fitness Standards
MPV	Military Pattern Vehicle
MSA	Mine Safety Appliances
MXF	Medium Expansion Foam
NAD	Naval Armament Detachment
NAFVN	North American Firefighter Veterans Network
NATO/SC	NATO Standardization Council
NATO	North Atlantic Treaty Organization
NCM	Non-Commissioned Members
NCO	Non-Commissioned Officers
NDHQ	National Defence Headquarters
NFPA	National Fire Protection Association
NLZ	No Lone Zone
NORAD	North American Aerospace Defense
NRL	US Navy Research Laboratories
NWSR	North West Staging Route
NYFD	City of New York Fire Department
NYPD	City of New York Police Department
OBE	Order of the British Empire
OC	Officer Commanding
OC FFS	Officer Commanding Firefighting School
OJTQS	On-Job Trade Qualification Standard
OMM	Order of Military Merit
OP EVAL	Operations Evaluation
OS&Y	Outside Screw and Yoke
OSCAR	On-Scene Commanders
PASS	Personal Alert Safety System
PCA	Practical Critical Area
PERI	Physical Education and Recreation Instructors

PF	Protein Foam
PFL	Power Failure Landing
PK	Potassium Bicarbonate
PMO	Prime Minister's Office
PMQ	Permanent Married Quarters
PSC	Public Service Commission
PT	Physical Fitness Training
PX	Post Exchange
QHM	Queen's Harbour Master
QL	Trade Qualification Level
RAF	Royal Air Force
RCAF	Royal Canadian Air Force
RCASC	Royal Canadian Army Service Corps
RCC	Rescue Co-ordination Centre
RCE	Royal Canadian Engineers
RCEO	Regional Construction Engineering Office
RCHA	Royal Canadian Horse Artillery
RCN	Royal Canadian Navy
RCNVR	Royal Canadian Navy Volunteer Reserve
RFC	Royal Flying Corps
RFM	Regional Fire Marshall
RIV	Rapid Intervention Vehicle
RMC	Royal Military College
RNAS	Royal Naval Air Service
RPM	Revolutions Per Minute
RRA	Rapid Response Area
SADFD	South Area Ammunition Depot Fire Department
SAS	Special Armament Storage
SB	Substantive
SCBA	Self-Contained Breathing Apparatus
SCP	Special Commissioning Plan
SITREP	Situation Reports
SOP	Standard Operating Procedures
SRCP	Special Ranks Commissioning Plan
TAC EVAL	Tactical Evaluations
TCA	Theoretical Critical Area

TCP	Triple-Combination Pumper
TIC	Thermal-Imaging Camera
UCR	Unsatisfactory Condition Report
UK	United Kingdom
ULC	Underwriters' Laboratories of Canada
US	United States
USAF	United States Air Force
UTPM	University Training Plan Men
UTPNCM	University Training Plan, Non-Commissioned Members
WFD	Wing Fire Department
WSFPC	War Services Fire Prevention Committee

APPENDIX D

HISTORY TEAMS

The History Teams were assembled and assigned an area of prime interest which, at least to some extent, was in keeping with each member's background. There was no intent to limit any team from taking a broader view; in fact, each was urged to do so.

Team Alpha (NDHQ)
Gaetan Perron
Peter Keough
Jack Henderson
Sean Tracey

Team Bravo (Navy)
Andy Beaudin
Jim McInnis
Ken Hoffer
Bob Connors

Team Charlie (Army)
Jeff Carlisle
Duff Rinehart
(Two others named to this team declined to serve.)

Team Delta (Air Force)
Hal Singleton
Garry Mauch
Jim Lockhart
Gaetan Morinville
Al Rau

Team Echo (CFE)
Bob Giguere
Lloyd Howlett
Dave Geddes
Charlie McNeil
Ian Morrison

Team Foxtrot (CFFA)
Bob Maxwell
Barry College
Joe Walker
Gary Oliver

Team Golf (Personnel)
Bob Morrison
Bob Maxwell
John Daley

Team Hotel (West Coast)
Charlie McNeil
Paul Beaulieu

Team India (Comments re: Cowan/Evans papers)
Ev Evans

Team Juliet (Decorations & Awards)
Jim Munro
John Daley

Team Kilo (CAFC & ACFM/FC)
Marcel Ethier
Larry Lamarches

Team Lima (Apparatus & Materiel)
Don Carmichael
Frank O'Meara
Jack Kroeplin
Jim Wright

Team Mike (Illustrations/Photos)
Steve Shand
Archie Graham
Don McNeil

Team Oscar (Editorial Group)
Lorne MacLean
Don Carmichael
Dough Stevenson
Mike Blow
Andy Beaudin

.

Appendix E

Radar Stations

Alsask, Saskatchewan	1961–August 1986
Armstrong, Ontario	1952–October 1974
Baldy Hughes, British Columbia	1952–April 1988
Barrington, Nova Scotia	1956–August 1990
Beausejour, Manitoba	1953–July 31, 1986
Beaverbank, Nova Scotia	1954–April 1964
Beaverlodge, Alberta	1953–April 1988
Cartwright, Labrador	1951–June 1968
Chibougamau, Quebec	1961–April 1988
Cold Lake, Alberta	1954–August 1991
Comox, British Columbia	1954–June 1958
Dana, Saskatchewan	1961–1987
Edgar, Ontario	1950–March 1964
Falconbridge, Ontario	1950–November 1986
Foymount, Ontario	1950–October 1974
Frobisher Bay, Northwest Territories	1951–November 1961
Gander, Newfoundland	1952–July 1990
Goose Bay (Melville), Labrador	1951–July 1988
Gypsumville, Manitoba	1961–July 1987
Holberg, British Columbia	1950–January 1991
Hopedale, Labrador	1951–June 1968
Kamloops, British Columbia	1956–April 1988
Lac St. Denis, Quebec	1949–August 1986
Lowther, Ontario	1956–April 1987
Moisie, Quebec	1951–August 1988
Mont Apica, Quebec	1951–August 1990
Moosonee, Ontario	1961–August 1975
Pagwa, Ontario	1951–October 1966
Parent, Quebec	1951–April 1964
Penhold, Alberta	1961–August 1986
Puntzi Mountain, British Columbia	1950–October 1966
Ramore, Ontario	1950–September 1974
Resolution Island, Northwest Territories	1951–November 1961

Saglek, Labrador	1951–June 30, 1970
Senneterre, Quebec	1950–August 1988
Sioux Lookout, Ontario	1951–July 1987
St. Anthony, Newfoundland	1951–March 1968
Stephenville, Newfoundland	1951–1971
St. John's (Red Cliff), Newfoundland	1951–October 1961
St. Margaret's, New Brunswick	1950–April 1988
St. Sylvestre, Quebec	1952–April 1964
Sydney, Nova Scotia	1954–January 1991
Tofino, British Columbia	1952–December 1957
Yorkton, Saskatchewan	1961–August 1986

APPENDIX F

FIREFIGHTERS—MINIMUM ARFF STANDARDS

STANAG 7145—Minimum Core Competency Levels and Proficiency of Skills for NATO Firefighters

Excerpt
6. The proficiency of aircraft CFR skills requires, as a minimum, training of all firefighters at the following frequencies:

a. Crew extraction on each assigned aircraft including shutdown, safety procedures, and personnel extraction/rescue semi-annually, unless more frequently is determined by the fire chief. As a prerequisite for aircraft equipped with pyrotechnic escape systems and devices, personnel must have cockpit/flight deck familiarization training prior to aircraft entry. This training is normally accomplished by escape system technicians. It is recommended that an annual refresher be accomplished to keep fire fighters abreast of any changes to the escape system.

b. Live training fires during:
   i. daytime—no less than once every 6 months
   ii. darkness—no less than once per year.

7. It is becoming increasingly difficult to maintain a live fire training facility that is environmentally acceptable while still providing realistic training. Consequently, authentic training aids and methods may have to be devised to meet the requirements of this STANAG.

NFPA 405
Recommended Practice for the Recurring Proficiency Training of Aircraft Rescue and Fire-Fighting Services. 1999 Edition.

Excerpt

12-1 General.
This chapter identifies the various types and sizes of fires associated with aircraft accidents. ARFF personnel should regularly demonstrate, individually and as teams, their ability to safely and effectively control and extinguish these fires.

12-2 Criteria.

12-2.1

ARFF personnel should be able to extinguish the following:

(1) An aircraft fuel fire or simulation utilizing an appropriate fire extinguisher.
(2) An aircraft fuel fire or simulation utilizing ARFF vehicle hand lines and appropriate extinguishing agent while using proper technique. The size of the fire should be appropriate for the agent flow of the hand lines.
(3) An aircraft fuel fire or simulation using ARFF vehicle turrets and appropriate extinguishing agent and proper technique. The size of the fire should be appropriate for the class or index of the airport.
(4) A simulated three-dimensional fire, using ARFF vehicle hand lines, appropriate extinguishing agent(s), and proper technique.
(5) A simulated aircraft cabin fire, using ARFF vehicle hand lines and water spray. Hand lines should be properly advanced and co-ordinated with ventilation operations.
(6) A simulated auxiliary power unit fire on an aircraft, utilizing ARFF vehicle hand lines or turrets that apply appropriate extinguishing agent and using proper technique.
(7) A simulated aircraft wheel/brake area fire, utilizing an ARFF vehicle hand line and appropriate agent and proper technique.
(8) A simulated electrical fire, utilizing the appropriate extinguishing agent and proper procedures and technique.
(9) A simulated engine fire, using an ARFF vehicle hand line and appropriate extinguishing agent and the proper technique.

12-2.2

Measurable standards of highly skilled performance should be established for each of the fire situations listed in 12-2.1, (1) through (9), by the authority having jurisdiction for the airport.

APPENDIX G

CALCULATING MINIMUM ARFF REQUIREMENTS

Up until the 1970s, crash protection, including agent quantities and number of vehicles required at airports, was based on the fuel load and passenger capacity of the aircraft using the airport. The NFPA 403[1] Committee had been gathering statistical data from hundreds of crashes all over the world and developed new criterion for calculating ARFF requirements.

The first step in the process of ARFF requirements at a given airport was to establish a method of categorizing the risk based on its traffic patterns. Recognizing that the size of the aircraft involved was a key factor, the overall length and width of aircraft operating in and out of a given airport was used to establish a scale of risk categories ranging from Category 1 up to Category 10. This was done based on analysis of data collected on aircraft accidents worldwide and on the results of extensive controlled testing.

The military added a caveat that ARFF services would be based either on the size of the aircraft or on its fuel capacity, whichever was the greater, in order to cover for the fact that certain types of military aircraft carry a greater fuel load than a commercial aircraft of similar size.

Table H-1
Airport Category by Overall Length, Width and Fuel Capacity of Aircraft

			Overall Length (up to but not including)		Overall Width (up to but not including)		Maximum Fuel Capacity	
NFPA	FAA	ICAO	ft	m	ft	m	L	G
1	GA-1	1	30	9	6.6	2	400	90
2	GA-1	2	39	12	6.6	2	1000	220
3	GA-2	3	59	18	9.8	3	2500	550
4	A	4	78	24	13.0	4	6250	1400
5	A	5	90	28	13.0	4	15500	3400
6	B	6	126	39	16.4	5	40000	8800
7	C	7	160	49	16.4	5	100000	22000
8	D	8	200	61	23.0	7	200000	44000
9	E	9	250	76	23.0	7	400000	88000
10			300	91	25.0	8	400000+	88000+

---

[1] NFPA 403, Standard for Aircraft Rescue and Fire-Fighting Services.

Minimum quantities of extinguishing agents, minimum agent application rates, along with the number of vehicles, could then be established for each category, using criteria based on the Area Concept which includes three main factors:

a. The TCA, which is the area adjacent to the aircraft where the fire must be controlled for the purpose of ensuring temporary fuselage integrity and providing an escape area for its occupants;
b. The Practical Critical Area (PCA), which is two thirds of the TCA; and
c. The Control Time (CT), which is the time required from the arrival of the first firefighting vehicle to the time the initial intensity of the fire is reduced by 90%. Empirical data from a variety of sources suggests that one minute is a reasonable time to accomplish this task.

Calculating Quantities of Extinguishing Agents

The quantity of water for foam production required to reduce fire intensity in the PCA by 90% within one minute is referred to as Q1. In turn, Q2 relates to the need to have sufficient fire suppression agents available to maintain conditions that do not pose a threat to life in the PCA until such time as rescue operations are completed. The secondary role of Q2 is to extinguish all fires in and peripheral to the PCA. Q3 is the water required for interior firefighting using hand lines.

Data collected over the years now permits us to specify the required application rates for three generic foam types needed to extinguish fire in $1\ m^2 / 1\ ft^2$ of the PCA as follows:

(a) AFFF = 5.5 L/min/m^2 / 0.13 gpm/ft^2
(b) FP = 7.5 L/min/m^2 / 0.18 gpm/ft^2
(c) PF = 8.2 L/min/m^2 / 0.20 gpm/ft^2

The method for calculating the values for each component of Q are presented below.

Where PCA = (0.67) TCA, TCA = L (K + W), and
L = length of aircraft
W = width of fuselage
R = application rate of selected agent
T = time of application (1 minute)
K = values shown below

K — Feet
K= 39 where L = less than 39
K= 46 where L = 39 up to but not including 59
K= 56 where L = 59 up to but not including 79
K= 98 where L = 79 and over

K — Metres
K = 12 where L = less than 12
K = 14 where L = 12 up to but not including 18
K = 17 where L = 18 up to but not including 24
K = 30 where L = 24 and over

Table H-2
Q2 as a Percentage of Q1

Airport Category	Q2 % of Q1	Airport Category	Q2 % of Q1
1	0	6	100
2	27	7	129
3	30	8	152
4	58	9	170
5	75	10	190

The values of Q3 are based on accepted water flow requirements for the type of fire-fighting operations to be experienced when combatting an interior aircraft fire. They are determined as follows:

Table H-3
Quantity of Water and Discharge Rate for Q3

Airport Category	Q3 — Litres		Airport Category	Q3 — Litres	
1	0	0	6	35 L/min	10 min = 4750 L
2	0	0	7	35 L/min	10 min = 4750 L
3	20 L/min	5 min = 1135 L	8	70 L/min	10 min = 9500 L
4	20 L/min	10 min = 2270 L	9	70 L/min	10 min = 9500 L
5	35 L/min	10 min = 4750 L	10	70 L/min	10 min = 9500 L

A sample calculation using Airport Category 7 based on aircraft measuring 48.99 m in length and 4.99 m in width and AFFF as the primary extinguishing agent is shown below.

TCA – 48.99m X (4.99 m + 30-K factor) = 1714.16 m^2 of TCA

PCA – 2/3 X 1714.16 m = 1142.76 m^2 of PCA

Q1 – 5.5 L/M m^2 X 1142.67 m^2 = 6285 L water

Q2 – 129% of Q1 (129% X 1142.76L) = 8108 L water

Q3 – 3697 L water
Q – 6285 L/Q1 + 8108 L/ Q2 + 4732 L/ Q3 = 19125 L water
Q Rounded 19,000 L water

The authority having jurisdiction shall determine the level of protection based on the largest aircraft scheduled into the airport. Airports shall be categorized for ARFF services in accordance with Table H-1.

Table H-4
Minimum Extinguishing Agent Quantities and Discharge Rates

	AFFF		FP/FFFP		PF		PK		Halon 1211*	
	W	DR/L	W	DR/L	W	DR/L	kg	DR/K	kg	DR/K
1	447	225	617	290	685	322	45	2.25	45	2.25
2	738	500	1022	680	1136	738	90	2.25	90	2.25
3	2535	800	3073	1165	3289	1268	135	2.25	135	2.25
4	5052	1500	6124	2000	6552	2176	135	2.25	135	2.25
5	10454	3000	12657	4300	13531	4656	205	2.25	205	2.25
6	14171	4000	17805	5600	19254	6132	205	2.25	205	2.25
7	18459	5500	23740	7450	25852	8138	205	2.25	205	2.25
8	29440	7000	37123	9850	40197	10768	410	4.5	410	4.5
9	36222	9000	46514	12500	50632	13172	410	4.5	410	4.5
10	44527	11700	58009	15500	63406	17411	410	4.5	410	4.5

* Or other approved Halon alternate

W – Litres of water
kg – kilograms PK
DR/L– Discharge rate, L/min
DR/K – Discharge rate, kg/min

Appendix H

Operations Assignments

Home Unit	Surname	Tasking Name	Location	Start Date	End Date
22 WG North Bay	HORNER	LFCCA	CYPRUS	10-01-1984	03-01-1985
3 WG Bagotville	DUPRAS	CAREER MANAGER TASKING	CYPRUS	26-02-1985	01-09-1985
19 WG Comox	DION	1 CAD TASKING	MIDDLE EAST	06-06-1986	15-12-1986
4 WG Cold Lake	JOHNMAN	CAREER MANAGER TASKING	CYPRUS	01-09-1986	01-03-1987
22 WG North Bay	RIDDELL	CAREER MANAGER TASKING	CYPRUS	01-03-1989	01-09-1989
Ex CFB Ottawa	ALBERT	CAREER MANAGER TASKING	MIDDLE EAST (Syria)	30-05-1989	13-09-1989
14 WG Greenwood	NANTEL	CAREER MANAGER TASKING	MIDDLE EAST	06-03-1990	12-09-1990
14 WG Greenwood	JENNINGS	OP FRICTION	KUWAIT	17-07-1990	24-08-1990
17 WG Winnipeg	MATTHEWS	OP FRICTION	KUWAIT	24-08-1990	10-01-1991
4 WG Cold Lake	THIBEAULT	OP FRICTION	KUWAIT	24-08-1990	07-04-1991
8 WG Trenton	BRESKE	OP FRICTION	KUWAIT	11-09-1990	10-02-1991
4 AES Cold Lake	COISH	OP FRICTION	KUWAIT	11-11-1990	19-02-1991
4 WG Cold Lake	OROURKE	OP FRICTION	KUWAIT	26-11-1990	14-03-1991
4 WG Cold Lake	CASSELMAN	OP FRICTION	KUWAIT	04-01-1991	02-08-1991
19 WG Comox	CYR	OP FRICTION	KUWAIT	04-01-1991	02-08-1991
Edmonton	VEZEAU	OP FRICTION	KUWAIT	05-02-1991	12-04-1991
14 WG Greenwood	BABINEAU	OP FRICTION	KUWAIT	21-08-1991	06-04-1992
22 WG North Bay	FIELDING	CAREER MANAGER TASKING	MIDDLE EAST	01-09-1992	10-03-1993
3 WG Bagotville	BOUDREAU	OP DANACA	MIDDLE EAST	02-03-1993	01-09-1993
4 WG Cold Lake	LEONARD	OP CAVALIER	YUGOSLAVIA	25-09-1993	25-04-1994
22 North Bay	HORNER	OP TEMPEST	HAITI	16-10-1993	23-11-1993
4 WG Cold Lake	FORGET	OP MANDARIN	YUGOSLAVIA	15-01-1994	15-06-1994
4 WG Cold Lake	GENTILE	EX WINGED BEAVER 9401	HAWAII	11-02-1994	27-02-1994
Ex CFB Ottawa	GRONDINE	OP CAVALIER	YUGOSLAVIA	02-04-1994	12-09-1994
19 Comox	HAMILTON	OP MANDARIN	YUGOSLAVIA	15-04-1994	15-09-1994
14 WG Greenwood	FARRELL	AIRCOM TASKING	MIDDLE EAST	28-08-1994	27-02-1995
22 WG North Bay	HORNER	1 CAD TASKING	YUGOSLAVIA	02-01-1995	06-01-1995
4 WG Cold Lake	OCHITWA	OP MANDARIN	YUGOSLAVIA	16-02-1995	09-09-1995
3 WG Bagotville	BELANGER	AIRCOM TASKING	MIDDLE EAST	27-02-1995	05-09-1995
19 Comox	DESRUISSEAUX	OP PIVOT	HAITI	15-03-1995	15-09-1995
Ex CFB Ottawa	RUTHERFORD	OP PIVOT	HAITI	24-03-1995	24-09-1995
8 WG Trenton	COMEAU	OP PIVOT	HAITI	24-03-1995	24-09-1995
17 WG Winnipeg	PEPIOT	OP PIVOT	HAITI	26-03-1995	26-09-1995
Ex CFB Shearwater	GAUTHIER	MOBILE ARRESTOR GEAR	GRAND PRAIRIE,	02-08-1995	08-08-1995
8 WG Trenton	MACINTYRE	MOBILE ARRESTOR GEAR	KAMLOOPS, BC	05-08-1995	11-08-1995

Home Unit	Surname	Tasking Name	Location	Start Date	End Date
Ex CFB Shearwater	GELDART	MOBILE ARRESTOR GEAR	ABBOTSFORD, BC	07-08-1995	14-08-1995
Ex CFB Shearwater	GELDART	MOBILE ARRESTOR GEAR	LETHBRIDGE, AB	15-08-1995	21-08-1995
15 WG Moose Jaw	PEARSON	AIRCOM TASKING	MIDDLE EAST	28-08-1995	05-03-1996
3 WG Bagotville	DUGUAY	OP PIVOT	HAITI	31-08-1995	17-01-1996
15 WG Moose Jaw	CHARLAND	OP PIVOT	HAITI	20-09-1995	01-11-1995
8 WG Trenton	MEITZ	OP PIVOT	HAITI	20-09-1995	15-04-1996
Ex CFB Shearwater	LARIVIERE	OP PIVOT	HAITI	22-09-1995	15-04-1996
19 Comox	ST-PIERRE	OP PIVOT	HAITI	22-09-1995	15-04-1996
8 WG Trenton	BOLDUC	OP PIVOT	HAITI	24-09-1995	15-04-1996
4 WG Cold Lake	DUMOUCHEL	OP PIVOT	HAITI	24-09-1995	15-04-1996
4 WG Cold Lake	CASSELMAN	OP PIVOT	HAITI	24-09-1995	15-04-1996
19 Comox	SCOTT	OP PIVOT	HAITI	24-09-1995	15-04-1996
19 Comox	LAZAROV	OP PIVOT	HAITI	24-09-1995	15-04-1996
3 WG Bagotville	HUDON	OP PIVOT	HAITI	24-09-1995	15-04-1996
19 Comox	DATCHKO	OP PIVOT	HAITI	24-09-1995	15-04-1996
3 WG Bagotville	LEBLANC	OP PIVOT	HAITI	24-09-1995	17-01-1996
3 WG Bagotville	LANGLAIS	OP PIVOT	HAITI	24-09-1995	15-04-1996
14 WG Greenwood	MESSIER	OP PIVOT	HAITI	24-09-1995	15-04-1996
15 WG Moose Jaw	HOLLINS	OP PIVOT	HAITI	23-11-1995	15-04-1996
4 WG Cold Lake	DALRYMPLE	AIRCOM TASKING	MIDDLE EAST	26-02-1996	31-08-1996
3 WG Bagotville	TREMBLAY	OP STANDARD	HAITI	13-05-1996	13-11-1996
3 WG Bagotville	THIVIERGE	AIRCOM TASKING	MIDDLE EAST	27-08-1996	07-03-1997
14 WG Greenwood	BOULOS	OP STABLE	HAITI	01-10-1996	31-03-1997
8 WG Trenton	BLACKMORE	OP STABLE	HAITI	01-10-1996	31-03-1997
14 WG Greenwood	GRANT	OP STABLE	HAITI	01-04-1997	01-10-1997
14 WG Greenwood	HALLER	OP STABLE	HAITI	01-04-1997	01-10-1997
14 WG Greenwood	LECOMPTE	OP STABLE	HAITI	01-04-1997	01-10-1997
14 WG Greenwood	OATES	OP STABLE	HAITI	01-04-1997	01-10-1997
17 WG Winnipeg	HUNTRODS	OP PALLADIUM	YUGOSLAVIA	15-07-1997	15-01-1998
8 WG Trenton	TUCKER	OP PALLADIUM	YUGOSLAVIA	15-07-1997	15-01-1998
15 WG Moose Jaw	DIXON	AIRCOM TASKING	MIDDLE EAST	14-08-1997	04-03-1998
8 WG Trenton	VERCH	OP PALLADIUM	BOSNIA	01-01-1998	30-07-1998
4 WG Cold Lake	SMITH	OP PALLADIUM	YUGOSLAVIA	04-01-1998	20-07-1998
8 WG Trenton	MACNEIL	OP PALLADIUM	YUGOSLAVIA	04-01-1998	20-07-1998
15 WG Moose Jaw	THIVIERGE	OP PALLADIUM	YUGOSLAVIA	04-01-1998	20-07-1998
19 Comox	WATSON	OP PALLADIUM	SARAJEVO	05-06-1998	07-12-1998
8 WG Trenton	MACNEIL	OP PALLADIUM	YUGOSLAVIA	10-07-1998	20-01-1999
19 WG Comox	ALLEY	OP PALLADIUM	YUGOSLAVIA	10-07-1998	20-01-1999
4 WG Cold Lake	GORRIE	OP DANACA	MIDDLE EAST	19-08-1998	02-03-1999
19 Comox	WEATHERHEAD	OP PALLADIUM AIRDET	BOSNIA	01-10-1998	01-02-1999
19 Comox	CYR	OP PALLADIUM AIRDET	BOSNIA	01-10-1998	01-02-1999
19 Comox	PRIOR	OP PALLADIUM AIRDET	BOSNIA	01-10-1998	01-02-1999
19 Comox	CLOUTIER	OP PALLADIUM AIRDET	BOSNIA	01-10-1998	01-02-1999

Home Unit	Surname	Tasking Name	Location	Start Date	End Date
3 WG Bagotville	MADORE	OP PALLADIUM	BOSNIA	09-10-1998	21-11-1998
4 WG Cold Lake	JOLIN	OP PALLADIUM	BOSNIA	09-10-1998	21-11-1998
8 WG Trenton	ROBINSON	OP GUARANTOR	MACEDONIA	19-12-1998	19-03-1999
8 WG Trenton	ANDERSON	OP GUARANTOR	MACEDONIA	19-12-1998	19-03-1999
8 WG Trenton	BRESKE	OP GUARANTOR	MACEDONIA	19-12-1998	19-03-1999
8 WG Trenton	RUNNING	OP GUARANTOR	MACEDONIA	19-12-1998	19-03-1999
19 Comox	HUARD	1 CAD TASKING	BOSNIA	08-01-1999	24-01-1999
8 WG Trenton	HILLYARD	OP PALLADIUM	BOSNIA	15-01-1999	10-08-1999
14 WG Greenwood	NANTEL	OP PALLADIUM	BOSNIA	30-01-1999	30-07-1999
3 WG Bagotville	BELANGER	OP PALLADIUM AIRDET	BOSNIA	15-02-1999	15-08-1999
4 WG Cold Lake	PROULX	OP PALLADIUM AIRDET	BOSNIA	15-02-1999	15-08-1999
17 WG Winnipeg	THIBEAULT	OP PALLADIUM AIRDET	BOSNIA	15-02-1999	15-08-1999
3 WG Bagotville	GAUDETTE	OP PALLADIUM AIRDET	BOSNIA	15-02-1999	15-08-1999
19 Comox	GOODWIN	OP DANACA	MIDDLE EAST	01-03-1999	05-09-1999
14 WG Greenwood	PENNEY	OP KINETIC	KOSOVO	22-03-1999	23-12-1999
14 WG Greenwood	PLETSCH	OP KINETIC	KOSOVO	22-03-1999	23-12-1999
14 WG Greenwood	HALLER	OP KINETIC	KOSOVO	22-03-1999	23-12-1999
14 WG Greenwood	FIELDING	OP KINETIC	KOSOVO	22-03-1999	23-12-1999
14 WG Greenwood	DWYER	OP KINETIC	KOSOVO	22-03-1999	23-12-1999
4 WG Cold Lake	CASSELMAN	TAV AVIANO	AVIANO	30-04-1999	05-07-1999
4 WG Cold Lake	MCGREGOR	TAV AVIANO	AVIANO	30-04-1999	05-07-1999
3 WG Bagotville	GIROUARD	OP PALLADIUM AIRDET	BOSNIA	01-08-1999	28-02-2000
3 WG Bagotville	DESCOTEAUX	OP PALLADIUM AIRDET	BOSNIA	01-08-1999	28-02-2000
4 WG Cold Lake	SKIBINSKY	OP PALLADIUM AIRDET	BOSNIA	01-08-1999	28-02-2000
3 WG Bagotville	BLACKBURN	OP PALLADIUM	BOSNIA	10-08-1999	28-02-2000
3 WG Bagotville	FORGET	OP PALLADIUM	BOSNIA	10-08-1999	01-07-2003
8 WG Trenton	GARVIN	OP PALLADIUM AIRDET	BOSNIA	15-08-1999	28-02-2000
19 Comox	FORGET	OP PALLADIUM	BOSNIA	15-08-1999	15-12-1999
CFB Halifax	MARTIN	OP PALLADIUM	BOSNIA	15-08-1999	28-02-2000
3 WG Bagotville	LONGPRE	OP DANACA	MIDDLE EAST	25-08-1999	07-03-2000
4 WG Cold Lake	BROTHERTON	OP PALLADIUM	BOSNIA	01-11-1999	28-02-2000
8 WG Trenton	BARKER	OP KINETIC	KOSOVO	01-12-1999	10-04-2000
4 WG Cold Lake	ST-PIERRE	OP KINETIC	KOSOVO	10-12-1999	10-06-2000
4 WG Cold Lake	WAGG	OP KINETIC	KOSOVO	10-12-1999	10-06-2000
4 WG Cold Lake	PARADIS	OP KINETIC	KOSOVO	10-12-1999	10-06-2000
4 WG Cold Lake	JOLIN	OP KINETIC	KOSOVO	10-12-1999	10-06-2000
19 Comox	BETTESWORTH	OP KINETIC	KOSOVO	15-12-1999	15-06-2000
19 Comox	WEATHERHEAD	OP KINETIC	KOSOVO	15-12-1999	15-06-2000
14 WG Greenwood	FOWLER	OP PALLADIUM	BOSNIA	21-02-2000	31-08-2000
17 WG Winnipeg	PRICE	OP PALLADIUM	BOSNIA	21-02-2000	31-08-2000
19 Comox	SCOTT	OP PALLADIUM	BOSNIA	21-02-2000	31-08-2000
8 WG Trenton	SYLVESTER	OP PALLADIUM	BOSNIA	21-02-2000	25-09-2000
14 WG Greenwood	FREEMAN	OP PALLADIUM	BOSNIA	21-02-2000	25-09-2000

Home Unit	Surname	Tasking Name	Location	Start Date	End Date
3 WG Bagotville	PERUSSE	OP PALLADIUM	BOSNIA	21-02-2000	25-09-2000
8 WG Trenton	BRESKE	OP PALLADIUM	BOSNIA	21-02-2000	30-06-2000
3 WG Bagotville	LAFRENIERE	OP PALLADIUM	BOSNIA	21-02-2000	31-08-2000
14 WG Greenwood	VERRALL	OP DANACA	MIDDLE EAST	24-02-2000	06-09-2000
4 AES Cold Lake	LOWE	OP DANACA	MIDDLE EAST	22-08-2000	06-03-2001
3 WG Bagotville	ST-JACQUES	AVIANO CAMP TEAR-DOWN	AVIANO	09-03-2001	23-04-2001
3 WG Bagotville	GILCHRIST	AVIANO CAMP TEAR-DOWN	AVIANO	09-03-2001	23-04-2001
4 WG Cold Lake	GAUDREAU	OP DANACA	MIDDLE EAST	17-08-2001	04-03-2002
3 WG Bagotville	LAROCHE	OP PALLADIUM	CORALICI	11-09-2001	18-10-2001
8 WG Trenton	BARKER	OP APOLLO CC 130 VANGUARD	DUBAI UAE	21-01-2002	25-07-2002
19 Comox	SMITH	OP PALLADIUM	BOSNIA	15-02-2002	15-10-2002
3 WG Bagotville	LEBLANC	OP DANACA	MIDDLE EAST	27-02-2002	02-09-2002
4 WG Cold Lake	SKIBINSKY	SLTA	ALERT	13-03-2002	11-04-2002
8 WG Trenton	DUCHESNE	OP PALLADIUM	BOSNIA	15-03-2002	15-10-2002
4 WG Cold Lake	WAGG	OP PALLADIUM	BOSNIA	15-03-2002	15-10-2002
14 WG Greenwood	FRANCOEUR	OP PALLADIUM	BOSNIA	15-03-2002	15-10-2002
3 WG Bagotville	COTE	OP PALLADIUM	BOSNIA	15-03-2002	15-10-2002
3 WG Bagotville	TIMCHUCK	OP PALLADIUM	BOSNIA	15-03-2002	15-10-2002
14 WG Greenwood	REDDEN	OP PALLADIUM	BOSNIA	15-03-2002	15-10-2002
3 WG Bagotville	GROLEAU	OP APOLLO	AL-MINHAD U.A.E	01-04-2002	04-10-2002
14 WG Greenwood	BRUENS	OP DANACA	MIDDLE EAST	16-08-2002	03-03-2003
14 WG Greenwood	LONG	OP APOLLO	AL-MINHAD U.A.E	13-01-2003	18-07-2003
3 WG Bagotville	TREMBLAY	OP APOLLO (SHIP TOUR)	PERSIAN GULF	02-02-2003	01-07-2003
3 WG Bagotville	LONGPRE	OP APOLLO (SHIP TOUR)	PERSIAN GULF	02-02-2003	01-07-2003
19 Comox	PEPIOT	OP DANACA	MIDDLE EAST	19-02-2003	03-09-2003
4 WG Cold Lake	SIMMONDS	1 CAD TASKING	AL-MINHAD UAE	25-04-2003	27-05-2003
3 WG Bagotville	BOUCHARD	OP DANACA	MIDLE EAST	12-06-2003	08-07-2003
3 WG Bagotville	DESRUISSEAUX	OP ATHENA	AL-MINHAD UAE	23-07-2003	14-02-2004
4 WG Cold Lake	SIMMONDS	OP ATHENA	AL-MINHAD UAE	23-07-2003	14-02-2004
8 WG Trenton	BOUDREAU	OP ATHENA	AL-MINHAD UAE	23-07-2003	14-02-2004
14 WG Greenwood	FERRAR	OP ATHENA	AL-MINHAD UAE	23-07-2003	14-02-2004
19 Comox	AMOS	OP ATHENA	AL-MINHAD UAE	27-07-2003	13-02-2004
3 WG Bagotville	BOUCHARD	OP DANACA	MIDLE EAST	20-08-2003	03-03-2004
8 WG Trenton	PERUSSE	HLTA	MIDDLE EAST	03-01-2004	05-02-2004
17 WG Winnipeg	DATCHKO	OP ATHENA	AL-MINHAD UAE	21-01-2004	07-09-2004
19 Comox	GIRARD	TAV OP ATHENA	AL-MINHAD UAE	22-01-2004	20-02-2004
17 WG Winnipeg	MCGORY	TAV OP ATHENA	AL-MINHAD UAE	23-01-2004	19-03-2004
4 WG Cold Lake	LOWE	OP ATHENA	AL-MINHAD UAE	17-02-2004	06-07-2004
8 WG Trenton	PERUSSE	OP DANACA	MIDDLE EAST	18-02-2004	01-09-2004
4 WG Cold Lake	LABARRE	OP HALO	HAITI	20-03-2004	15-09-2004
8 WG Trenton	LAREAU	OP ATHENA	AL-MINHAD UAE	06-08-2004	16-02-2005
19 Comox	CORNELIUS	OP DANACA	MIDDLE EAST	18-08-2004	02-03-2005

AUTOGRAPHS & NOTES

Autographs & Notes

Autographs & Notes

Autographs & Notes

Autographs & Notes

Autographs & Notes

Autographs & Notes

Autographs & Notes

Autographs & Notes

Autographs & Notes

Autographs & Notes

Autographs & Notes

Autographs & Notes

Autographs & Notes

Autographs & Notes

Autographs & Notes